War and Rumours of War: The Evidential Base for the Recognition of Warfare in Prehistory

Julie Wileman

BAR International Series 1984
2009

Published in 2016 by
BAR Publishing, Oxford

BAR International Series 1984

War and Rumours of War: The Evidential Base for the Recognition of Warfare in Prehistory

ISBN 978 1 4073 0516 5

BAR Publishing is the trading name of British Archaeological Reports (Oxford) Ltd. British Archaeological Reports was first incorporated in 1974 to publish the BAR Series, International and British. In 1992 Hadrian Books Ltd became part of the BAR group. This volume was originally published by Archaeopress in conjunction with British Archaeological Reports (Oxford) Ltd / Hadrian Books Ltd, the Series principal publisher, in 2009. This present volume is published by BAR Publishing, 2016.

Printed in England

BAR
PUBLISHING

BAR titles are available from:

BAR Publishing
122 Banbury Rd, Oxford, OX2 7BP, UK
EMAIL info@barpublishing.com
PHONE +44 (0)1865 310431
FAX +44 (0)1865 316916
www.barpublishing.com

Table of Contents

List of Figures

List of Tables

Introduction

Aims

The goal of this study is to examine the potential for the understanding and recognition of the processes and occurrence of prehistoric warfare through the development of a series of correlates, resulting in testable models that can be applied to the archaeological record.

Such models need to be flexible and applicable across different periods and in a variety of geographical areas. To this purpose, examples of evidence are included from a wide spectrum of sources. However, after offering definitions of warfare and considering the nature of its archaeological evidence, the correlates and models will, for comparative purposes, be applied to a number of case studies which are located in later prehistoric societies.

It is anticipated that the employment of correlates may result in the highlighting of types of evidence not usually cited in the discussion about prehistoric war, its forms and effects; however, there is recognition that this approach cannot offer proof of the occurrence of war. The aim is to encourage further debate on the range of potential evidence and its value in this sphere of archaeological research.

The Problem of War

Despite the numerous speculations over the nature of war and peace in early communities, what has been lacking is sufficient evidence, derived from archaeology, for the real levels and influences of conflict in the past. In the nineteenth and early twentieth centuries, there was, among many people, a Hobbesian assumption that prehistoric peoples were, if not constantly dripping with gore, at least generally prepared for war. A smaller, perhaps more idealistic, number of writers followed Rousseau in imagining a more golden, even communistic, early Paradise of simple, peaceful societies who only gradually fell from bliss, often as a result of the influence of technology.

The identification of the occurrence and results of warfare in the archaeological records of prehistoric societies has always been fraught with difficulties, allowing the possibilities both of denial that more than a few conflicts took place and, possibly, of overestimation of scales of violence. It is clear from the review offered in Chapter Two of the major forms of evidence usually cited that there is little or no general consensus about the meanings they convey for the understanding of past warfare. Nearly every instance of cited evidence can be, and has been, explained away in terms of crime, status behaviour, post-depositional effect, accident or some other causal factor. Several writers have called for the development of a series of archaeological correlates of warfare in order to address these problems (e.g. Ember & Ember, 1997).

This study seeks to provide some models for investigation, to suggest some areas for research and data-gathering, to highlight potentials and problems for the interpretation of evidence and to provide some frameworks for future appreciations of the concept of prehistoric war. If evidence can be sought and recognised for warfare on more extended scales, it may be possible to approach the questions of the prevalence, scale and influence of conflict on the development of societies with a little more certainty.

This investigation will generally confine itself to a discussion of physical evidence, but it is recognised that a human activity such as war will be inextricably connected to causal, procedural and concomitant factors resulting from psychological, political, cognitive, emotional, religious and charismatic behaviours. It is intended to leave discussion of these elements to those with greater expertise and understanding of the issues involved - it is with the possibilities offered by physical evidence recoverable by excavation, analysis and survey that this paper is concerned. These possibilities may then assist the development of cognitive and psychological explanations.

It is recognised that no one form of evidence alone can be accepted as proof of prehistoric warfare, and, even in combination, much data cannot be regarded as conclusive. It is contended that data and combinations of data require putting into context to achieve useful meaning - it is a prerequisite that there should be an understanding of the forms, scales and characteristics of the conflict type to which the evidence may be applied. This is not simple - human social relations are complex enough on their own, and in thinking about war we must factor in environmental, topographic, climate-related, artefactual, ritual and a host of other changes and relationships as well. Some general themes and sets of relationships may, however, be identifiable and offer outline models for use.

It may, therefore, be useful to consider the phenomenon of war in more general terms. Whilst violence can often be momentary and ephemeral, war or sustained episodes of violent behaviour are much longer term events. Wars have causes, which may be multiple, and which may build up over generations to the point of actual violence. They have symptoms, many of which may appear before weapons are drawn in earnest. They have modes of conduct, strategies and tactics, which may include extended periods of low-level attrition as well as large 'set-piece' battles. And above all, they have effects which may occur both over wide areas of land and through long temporal perspectives, far beyond the battlefield and the moment of violence itself.

The need to consider warfare as a process, rather than as an event, has been noted by several writers. Vencl maintained that "archaeology suffers from a general tendency to limit the consideration of war to the occurrence of a few static attributes.." and so "prehistoric warfare has not been conceptualised as a historical phenomenon with its social connections, causes, and consequences" (1984:118). Keeley (1996:114) makes this point with regard to the conditions leading to conflict when he considers the motivation behind prehistoric conflict, and Monks reiterates Vencl's position, noting that "evidence for warfare as a process rather than an event is rarely sought" (1997:4). The benefits of adopting a wider, more long-term archaeological approach to the study of warfare are encapsulated by Haas in his call for the study of long-term evolutionary patterns of social conflict through an explicit focus on material conditions reasonably reflected in the archaeological record (2001a:331).

In addition to a consideration of the evidence associated with the appearance and prosecution of war, it is possible that a useful field of study would be of the aftermaths of wars. Very few discussions of the *effects* of warfare have been undertaken. If war is to be studied properly by archaeology, we need to be able to answer the question "What happened next?". It is possible that in attempting to identify the *effects* of warfare in some sort of organised way, we may be in a better position to recognise its *occurrence, participants* and *frequency* in the prehistoric past. The combination of these forms of information might put us somewhere along the road towards achieving the usefulness to the wider study of social development that Vencl suggested archaeology could provide.

The present exercise will not be able to answer most of the questions raised, but it is hoped that it will be able to provide a possible approach to the creation of a methodological context, and will form the basis for further discussion of the issues surrounding this topic. It is recognised that the use of correlates may give rise to the dangers of self-fulfilling prophecy; any attempt at interpretation in archaeology is similarly at risk, and use of the correlates must be approached with the necessary and appropriate rigour and objectivity.

Part One - The Process of War

Chapter One

Levels of Conflict

Investigations have shown that wars often result from escalation of hostility from one level to another, depending on levels of stress, perceived threat, need or reward, desire for revenge or justice, political, religious or social imperatives and local conditions and changes. Thus each set of relationships must be regarded as interrelated with the others, and contributory to each level of increasing scale and duration.

The terminology used in discussions of prehistoric warfare tends to the imprecise, particularly in reference to scales of conflict. An attempt will be made in this chapter to identify the levels of hostility appropriate to the discussion, and to define the various implications of words such as war and warfare, conflict, raid, hostilities, battles and ambushes.

Since the beginning of the nineteenth century, we have tended to think of war in terms of mass armies, professional training and wide degrees of social and geographical involvement. In prehistoric societies, these scales of discussion are probably inappropriate. Some writers have, on this basis, denied that prehistoric conflict can be categorised as war per se. Indeed, Keeley (1996:10-14) has traced the origin of modern concepts of primitive war to a predisposition to consider warfare to be restricted, because of its complexity, to so-called 'civilized' societies. Because the number of casualties was likely to have been relatively low in 'primitive' war, compared to those of the nineteenth and twentieth centuries, prehistoric conflict has been viewed as 'relatively benign' (ibid:15). Nonetheless, for those on the receiving end of group violence, it hardly matters whether the enemy is numbered in tens or in thousands, and whether the arena of war comprises continents or a few villages. The individual effects are often the same; the proportion of casualties in 'primitive' conflicts may even be greater. Basil Greenhill, in trying to define the difference between ships and boats, concluded that "a boat is that which various people, at various times, have called a boat" (Greenhill, 1995:9), and for the purposes of this discussion, war is that which various people, at various times and in various places, may have experienced as war.

The words 'war' and 'warfare' are in themselves problematic. Various readings of their meanings are implicit in the literature of the subject, although rarely defined. Attempts to separate concepts of 'a war', 'a state of war' or 'generalised warfare' in prehistory cannot be achieved with success in the present state of our understanding of these phenomena, and indeed, may be unhelpful, providing, as they may, limiting assumptions about such conflicts. How people at the time experienced or conceptualised conflict is unknowable, but the nature of that experience is likely to have been highly variable, depending on the level, form and aims of the conflict through which they have lived, and whether they were on the winning or the losing side.

Attempts have been made to classify violence, including warfare, by scale. In an archaeological regard, those forms identified at the smallest end include conflict within families and between neighbouring residents, criminal acts and acts such as wife-beating or other forms of familial abuse too ephemeral to be recognised. So-called 'sport' violence between males within the group or between neighbouring groups and opportunistic violence committed by very small numbers of people among each other and their neighbours may account for some of the traces of violence noted in the archaeological record. These types of violence are well-represented in anthropological accounts (including Thorpe, 2006; Ferguson, 1997; Arnold, 1995 and Dent, 1983) and would be difficult to distinguish from the next largest scales without the acquisition of confirmatory or contradictory evidence from a number of sites over increasingly large territorial units.

Interpersonal violence may occur during war, and therefore evidence of such violence cannot be discounted out of hand in attempting to recognise warfare. In the same way, hostilities involving threatening, posturing, calling of insults, trespass and other transgressions may be part of the process or instigation of war; the word 'conflict' implies physical contact of an aggressive nature, which need not necessarily be part of a war, but could be. Clearly, the archaeologist would have difficulty in assessing whether evidence of isolated interpersonal violence is part of warfare without a great deal of corroborative material.

Raids, battles and ambushes are forms of conflict, and usually involve some form of violence, but are not necessarily warfare. However, raids, battles and ambushes may be utilised as strategies within a more general and aggressive pattern of hostile behaviour that we would recognise as war, based on its longer maintenance and wider influence.

At a slightly increased level from individual interpersonal violence is the kind of conflict that involves villages or kinship groups in a localised pattern of violence, often temporary in nature, but sometimes endemic, and therefore of a longer duration. An ambush or a single battle may not add up to a war, but repeated similar incidents between the same protagonists do. The kind of aggression noted at Crow Creek or Norris Farms in the USA (Milner, 1995, Milner et al, 1991; Zimmerman, 1997) may fall into this group; endemic aggression may be sporadic, and interspersed with periods of apparent peace, but its repeated and cumulative effects may allow us to categorise it as warfare. Low-grade endemic war may present particular difficulties for archaeological recognition because of its sometimes limited geographical scale and its tendency to be prosecuted intermittently.

Regionalised conflict involves the existence or creation of alliances between several communities, perhaps relatively informal but requiring maintenance and negotiation in order to oppose a common enemy. This level of violence may depend on the existence of chiefdom or tribal structures, because of the levels of organisation and control involved. War on this scale has, historically, been more readily recognised and studied in archaeology.

More complex societies can become involved in conflict on a greater and more complex scale, involving issues of status and wealth maintenance, external relations and trade, and they can become involved in territorial or imperialistic struggles that achieve the level of full-scale war as we would recognise it today. It is not intended to extend this discussion into the realms of full-scale modern warfare, with its technological, international and political characteristics, although some useful parallels may be drawn from some battles, campaigns or incidents.

These scales of conflict represent a continuum rather than a series of definable levels, and it seems clear that conflicts which begin in a very small way can escalate to much greater seriousness. The problem facing archaeology is to be able to differentiate between isolated incidents of violence and forms of war based on partial and incomplete evidence.

In considering war in prehistoric Northern Europe, the levels of activity with which we are concerned are likely to range from hostile relationships between individual communities and conflict between those groups, to struggles associated with the rise of larger tribal, political or economic units which may involve territorial alliances, control and expansion. To be able to differentiate between large-scale feuds and small-scale wars is probably impossible at this distance of time, and in any case, is probably not particularly helpful to the present discussion, which is concerned not with cultural phenomena but with physical evidence.

Causes, aims and precipitating factors of war

It is proposed that the phenomenon of war, and therefore the forms of evidence for it that may survive archaeologically, may be comprised initially of evidence associated with the causes of war, those prerequisites for violence identified through such disciplines as sociology, psychology, anthropology, climatology, environmental science, and the like. The prerequisites in themselves do not dictate that warfare will have taken place, but they indicate, in their presence, appropriate conditions for conflict. The problem of the causes of war has attracted a great deal of debate, much of it contradictory or non-specific. Identified factors include Malthusian analyses of the effects of population growth and density, migration and colonisation, psycho-social effects such as revenge, male dominance and ethnic differentiation, social-political concerns with the rise of complexity and status achievement, environmental stresses both real and perceived, sedentism, territoriality, economic exigencies, and the effects of technological advance and change in subsistence practices.

Causes of hostility should be separated from its perceived aims, and its precipitating factors. If war is to be contemplated, decisions need to be taken about the aims of resorting to violence - what did attackers hope to achieve? Were people aware of overarching environmental or social factors in their decision to take up arms, or were they responding to much smaller, more personal, needs within the general situation? Surely the latter, so the apparent, immediate, reasons for war may seem superficially different from, or only loosely related to, the underlying causal determinants. Archaeological evidence for people's individual motivations may be impossible to find, unless it can be discerned in further categories of study, specifically the areas of strategy and tactics of acts of war, and the various levels of effects and results of warfare. It is equally unlikely that archaeology can identify the spark that lit the tinder of individual conflicts or a series of hostilities - this is the prerogative of history. At best, we may be able to identify the approximate date and location of the first blows struck, but we could never posit the occurrence of an Archduke's assassination in Sarajevo to explain the First World War from archaeological evidence alone. The first blows, as that War demonstrates, may occur at a great distance from the precipitating act, and may depend on a multitude of other social, political, economic and emotional factors. The onset of war is a complex phenomenon.

Numerous attempts have been made to categorise the causes of war in simple societies. They have been variously designated as Darwinian, materialist or historical (Maschner & Reedy-Maschner, 1998) or psycho-social, economic and political (van Hooff, 1990), with further suggestions including natural disasters (Ember & Ember, 1997), and climatic and environmental change (Lambert, 1997; Ohlsson, 1999).

Darwinian and psycho-social causation

Darwinian and psycho-social explanations tend to base themselves in notions of natural aggression as instinctive in primate behaviour, or in discussions about the degree and course of adaptation of human social groups in response to change and stress. It has been suggested that the seeds of human warfare can be discerned early in the Palaeolithic, even that violent territorial competition between Homo sapiens sapiens and Homo neanderthalensis led to the extinction of the latter species by our ancestors (Pettit, 2000). Bigelow (1975:7) among others has seen violence as an essential aspect of the evolution of our species, and Corning (1975:369) lists a number of evolutionary advantages deriving from aggressive behaviour, including the development of larger brains, bipedalism, inventiveness, physical stamina and, while admitting the contribution of hunting, gathering and nurturing behaviour, concludes, provocatively, that "human evolution has been due, principally, to warfare between hominid groups", opinions not shared universally.

It is not proposed to follow these arguments in any detail here; whatever their relevance to questions about human social development, they refer to periods of time and forms of social grouping outside the scope of this discussion, as defined earlier.

In terms of the kinds of warfare found in Northern Europe from the Neolithic, the psycho-social motivations that may be relevant include the kinds of revenge or honour wars described by Chagnon (1983) and others, and comprising the following motivations:

- Family feuds, hereditary enmity
- as revenge for murder, manslaughter, etc.
- as retribution for infringement of sexual codes
- as punishment for trespasses
- as a penalty for black magic, witchcraft, etc.
- Machismo, prestige, status, honour, fame
- raids to obtain status symbols and insignia of power: head-hunting and scalping
- war as a part of rites of passage
- war as sport and excitement: an escape from boredom
- war as a safety valve for tensions originating inside the group
- Mystico-religious motives for the satisfaction of bloodthirsty gods, or to obtain victims for sacrifice
- ethnocentric-xenophobic aggression
- tribalistic delusions of superiority
- as revenge for sacrilege etc.

[after van Hooff (1990:50)].

Van der Dennen (accessed 2000) supports a number of these propositions, as does Vencl (1984). These types of motivations are likely to be particularly difficult to recognise archaeologically. If the majority of prehistoric war rested on these types of foundations, there is very little we could say about it in any concrete sense.

However, van Hooff goes on to consider a number of other motivational causes of aggression and usefully points out that a primitive warrior may be unable to recognise the link between the larger cause and the immediate motive for aggression, nor the connection between the larger cause and the effect of conflict (1990:51). Keeley (1996:114-115) also makes this point, in a discussion of the difficulties in the identification by anthropologists of real and precipitating causes of war, while Ohlsson (1999:76-77) breaks the process of war instigation into four factors: triggers - the events that lead people to choose violence over non-violence; targets - the perceived gains of the action; channels - the political, social, economic and national 'fault lines' along which sides are taken, and the catalysts - those factors which affect the speed, duration or intensity of the conflict. In such an analysis, the kinds of motivational factors listed above are likely to form the precipitating causes and catalysts of aggressive action. The overarching causes, perhaps unrecognised as such by the combatants themselves, may be factors that we can recognise in the archaeological record.

Demographic causation

Of these overarching motives, one of the most commonly discussed is that resulting from demographic pressure, either population growth or, more specifically, population density. The debate falls into two main groupings, those who support Malthusian conclusions about population pressures, and those who reject, or at least, modify them. These explanations are concerned with stress caused on societies by population increase which outstrips, or threatens to outstrip, the subsistence or growth potential of the community; population density increase which creates local subsistence and social stresses; and/or stress created by the appearance of immigrating groups, emigration of key sectors of the population, or imbalance in the constitution of the population.

The belief that increased numbers and/or density of occupation and settlement lead to social friction and eventually conflict is supported by Ember & Ember (1997) who suggest that population pressure is an important factor in external war, Lambert (1997) who suggests that friendly external relations become less necessary for finding mates or insuring food relief as the home group grows in size, Brothwell (1999) who notes the effects of both proximity and mobility of neighbouring groups, Mercer (1989a) who suggests that a surplus of labour is a factor in the development of war, and Vencl (1984) who cites low population density as a correlate of infrequency of war, among others. Questioning the proposition is Keeley, who notes that "groups with densities of less than one person per square mile are just as likely to engage in warfare each year as groups whose densities are hundreds of times higher" (1996:118) and that assessing comparative levels of population density stress in different environmental situations would be difficult. He also suggests that larger societies possess more mechanisms for avoiding conflict and relieving stress in other ways; Jochim (1981), while agreeing with the general principle that high population densities increase the chances of interpersonal conflict, crimes and competition, suggests that war actually creates population aggregation and its associated stresses as people group together for defence, take in refugees and abandon ruined farms. He rejects a similar correlation between population pressure and warfare as "environmental determinism" (1981:24).

Haas (2001b) rejects notions of war before settled agriculture simply on the grounds that populations had insufficient numerical potential to achieve the necessary levels of causation, but cites resource stress due to population growth as a causal factor of war after agriculture had been adopted.

If, however, we accept that there is a general agreement that increased population density may be a correlation of increased aggression, we have our first real opportunity to acquire confirmatory evidence for the incidence of prehistoric war. We will need to decide at what point the society in question would come under such stress as a result of increasing density, and what factors would aggravate social relations to the point of bloodshed. We would also need to decide whether the causes of violence were solely related to psychological stress from overcrowding, like rats in a maze [a point of view firmly rejected by Keeley (1996:118)] or whether we need to factor in the carrying capacity of the landscape, the forms of subsistence strategy and management, and socio-political characteristics of these societies.

Environmental causation

A popular explanation for the rise of circumstances leading to the development of warfare in prehistoric societies is the adoption of agriculture during the Neolithic. Carneiro (1978, 1992) suggests that the effect of war in the Palaeolithic was dispersive, while in the later Neolithic it became aggregative, related to the effect of creating bounded environments. Settled agriculture may provide a more reliable source of subsistence (although this is contested by some) but it brings with it certain costs; the investment of time and effort in preparing fields, sowing and maintaining crops, and building structures, as well as the necessity to create and protect storage facilities, which mean that people are no longer able to diversify their efforts, they become tied to the land and dependent upon place (O'Shea & Halstead, 1989; Ferguson 1997). Mercer (1989a) suggests that evidence from Northwest Europe reflects this view and traces the need for defence building to the Neolithic, and Haas (2001a) is clear that warfare first appears alongside agriculture and is closely associated with the problems created by increased sedentism, territoriality and concentration of resources. Jochim (1981) also believes warfare can be correlated with territoriality. Warfare on the frontier between Neolithic farmers and Mesolithic foragers has been suggested by Keeley (1997) to explain his (much disputed) identification of fortified villages in Belgium. The potentials for conflict include rival views of land ownership and exploitation, lack of choice and alternative strategies for subsistence farmers, and psychological factors including fear of disaster or shortages and envy. The appearance of settled agriculture may, therefore, be a correlate for the appearance of certain kinds of war; it does not help, however, with the recognition of war in other periods or for other motives.

External explanations of the onset of war include stresses upon subsistence or growth caused by climatic or environmental change, soil degradation due to these factors or to over-utilisation or bad management. These phenomena are perceived as offering threats, real or potential, to survival or maintenance of subsistence standards locally, to the prosecution of successful production and trade relations, and to the society's ability to compete successfully with its neighbours. Ember & Ember (1997) cite natural disasters as a causal factor for warfare, but in most accounts, it is subsistence stress that is usually claimed to create the circumstances for violent action between communities. Climatic change may be a factor (Lambert 1997) in some cases. Demographic growth may outstrip the ability of the subsistence economy (as suggested by many including Haas, 2001a; Ohlsson, 1999) or lead to depletion of the carrying capacity of the available resources. Ohlsson (ibid) also suggests that inequitable distribution of available resources creates sufficient scarcity to precipitate conflict. However, Maschner (1997) in his description of warfare on the Northwest coast of the United States, suggests that there was no evidence for resource stress in causation of hostilities in the region.

A number of writers have suggested that it is not necessarily actual resource shortage that causes the rise of violence, but rather the fear and uncertainty regarding possible shortages

(Ember & Ember 1997; Maschner & Reedy-Maschner, 1998).

A typology of environmentally-induced conflict has been created by the Swiss 'Environment and Conflicts Project' (ENCOP) (quoted in Ohlsson, 1999:46-47). This includes centre-periphery conflicts created by circumscription of rural populations in relation to central elites; ethno-politicised conflicts over resource distribution inequities "where ethnicity has been instrumentalised to serve either as a pattern of identification or as a ground for mobilisation" (ibid:46); regional migration conflicts caused by population displacement, or voluntary or forced migration related to resource acquisition; cross-boundary migration conflicts similarly motivated and stemming from poverty, underdevelopment, resource depletion and demographic rise, changes in land use and control and natural disasters; demographically induced conflicts related directly to rising demands on the carrying capacity of available arable land; conflicts over international waters and conflicts about global environmental issues. Some, at least, of these scenarios may be applicable in prehistoric contexts. The scale of social organisation is clearly a factor in the type of conflict identified in this list, as would be the nature of each society's external relations.

Ohlsson suggests that environmental pressure alone rarely causes warfare - a social component is required. "The adaptive capacity of the society in general, and that state in particular, is critical ....the environmental issue at hand must first be used for organising and mobilising affected groups along [fault-lines in society] and there must be both an opportunity and a clearly perceived advantage of taking to arms, for conflict to occur" (1999:48). The socio-political influences towards conflict will be considered in more detail below.

Environmental correlates for war may, therefore, include evidence for climatic or other change, soil depletion and/or degradation, and over-exploitation of natural resources. However, if the motivating factor for conflict is only the fear of shortage, archaeological evidence would be invisible. The relationships between environmental conditions and warfare are far from simple.

Socio-political causation

A commonly-cited cause of conflict is that created by levels of increasingly complex social stratification and organisation, leading to the emergence of chiefly castes who maintain control and status through the use of warrior elites and their ability to protect their peoples. Expanding neighbouring polities may also lead a society towards readiness for conflict; threats may come in the form of invasion, raiding, and commercial and industrial competitiveness.

War as a mechanism for the development of status and prestige has been suggested by Billman (1997) as an alternative to demographic explanations, and as an accompaniment to environmental explanations by Jochim (1981). Many writers have cited a connection between the desire of chiefs and elites to enhance their prestige and power through raiding and conquest, and the control of support through the distribution of booty and the claiming of tribute (including O'Shea & Halstead, 1989; Pantreau, 1989; van der Dennen & Falger, 1990; Maschner & Patton 1997; Monks, 1997; Maschner & Reedy-Maschner, 1998; Kristiansen & Rowlands, 1998) and some have cited warfare as a contributing factor in the development of social complexity, especially Carneiro (1970, 1978) who has claimed that "warfare is surely a prime mover in the origin of the state" (1970:734). This is contested by Webster (1975) who believes that warfare is a facilitator, but not a cause, of social complexity, a view shared by Claessen (2000), who sees war as a secondary factor, following socio-political, economic and ideological developments and Flannery (1972) who regards the evidence as being too ambiguous to be helpful either way.

Helbling (2006) suggests that warfare can develop as a result of the relatively anarchic structure of developing polities coupled with a degree of immobility created by competing neighbouring emergent unities. No one group, in these situations, can afford to pursue a unilateral strategy of peace as this could be interpreted as a sign of weakness. A security dilemma is created.

> *"Under conditions of mutual mistrust and the reciprocal threat of force, the survival of each group depends on its ability to become larger and stronger and to recruit more allies than its potential adversaries. But as the military superiority of one group inevitably entails the inferiority of the other groups, these other groups strive to become larger and to win more allies in turn".*
>
> (Helbling, 2006:123).

These efforts tend to place more stress on local resources, to support larger groups or to acquire goods to maintain prestige, attract allies or reward service, creating further uncertainty and pressure towards pre-emptive action (ibid).

Evidence of social complexity may, therefore, indicate a correlate of war, particularly if there is evidence of competing elites, but whether such evidence is a correlate of the causes of war, or perhaps of its effects, is unclear.

War may be regarded as a strategy for the purposes of social regulation - to vent tension and to provide a means of punishing transgressions (Vayda, 1968; Mercer, 1989a). It may be adopted as an alternative to the relief of stress through trading -

> *"Trade is most effective in alleviating limited deficits of specified items; as shortages become larger in magnitude and encompass a greater variety of goods, some alternative or additional strategy may be necessary. The redistribution of goods by either trade or raids requires that the goods be portable; when scarce goods are fixed, the redistribution of people through emigration or warfare is necessary"*
>
> (Jochim, 1981:194).

Ferguson suggests war could act as a problem-solving strategy:

> *"When more people within a regional system live in conditions of sedentism, concentrated material values, pronounced political hierarchies and boundaries, sharp ecological and demographic inequalities, and developed but unequally beneficial trade networks, it is more likely that some people, especially some leaders, will find themselves in situations where it appears reasonable to be violent. The more groups use violence regularly, the more readily will others do the same, or face physical or social elimination. Thus as a systemic process, war may become an established way of dealing with problems, and an entire region restructured for war"*
>
> (1997:339).

Complex Causation

"Confusion often arises between individual and collective motives, or among efficient, formal, and final causes" (Keeley, 1996:114). Not only this, but there may be a number of contributory factors leading to outbreaks of violence, and they may not occur contemporaneously, but be part of a cumulative sequence of events and actions. "The question of cause and effect is a complex, recalcitrant phenomenon" (Claessen, 2000:3) because it is clear that what seems to cause warfare in one situation fails to do so in another. The chain of determinism is so complicated by the myriad possible variables that it may, in fact, be misleading to adopt the simplistic approach to explanations by reference to the economic, socio-political, environmental, etc. along the lines of Quincy Wright (1965) and others. Nor can we assume that bellum omnium contra omnes is sufficient rationale. Haas (2001a:330) believes that "...archaeology comes to play a central and complementary role in the social sciences by helping us understand the causes and consequences of warfare and ethnic conflict across cultures and across time. It is archaeology and the archaeological record that hold the most appropriate intellectual resources to answer broad, pressing questions about the inevitable or intrinsic qualities of human warfare" but notes that the record is restrictive because it is incomplete; archaeology has particular difficulties with ideology and personal motivations.

Clearly, someone who is suffering because of environmental factors is likely to be more psychologically 'touchy' about perceived insults or unfairness. A warrior who has incidentally gained booty in a conflict about revenge may be motivated to continue hostilities for further booty even after that revenge is satisfied. Wars about religion or political ambition frequently contain causal elements about economics and envy.

Again, these scenarios offer little in the way of archaeological evidence potential. Overall, therefore, an examination of possible causal evidence shows that it is very limited and ambiguous.

Conclusion

It seems that the argument about what actually constitutes 'war' or 'warfare' is far from concluded; this paper will therefore consider a range of forms of conflict which involve the taking up arms of one community against another, but will generally exclude inter-community warfare and aggression as being, arguably, less well defined as 'war' by most writers in the field.

Consideration will be given to evidence for forms of causation likely to leave some trace in the archaeological record (environmental and demographic change, and forms of socio-political development and conflict), and will also attempt to consider inter-relationships between these, but will not concentrate on psycho-social or biological aggression, as this depends on forms of interpretation outside the scope of this discussion.

It is in the search for material evidence that this paper will concern itself; if wider forms of such evidence can be gathered, discussions about the motivations and psychology of past communities may be addressed with more confidence than at present.

Chapter Two - Physical Evidence of War

Secondary to the prerequisites or causes are the symptoms of war. These comprise actions taken by societies to prepare for aggression or defence, and many of these can be traced archaeologically. Again, the presence of symptoms does not serve as a proof that warfare has taken place, but can suggest that it's occurrence was viewed at least as a possibility by the people concerned.

This chapter will offer a review of a range of typical forms of cited evidence for warfare, drawn from descriptions of sites and artefacts derived from a wide cross-section of periods, places and cultures. Because warfare has a myriad of forms and effects that impact on human social, psychological and physical life, it may be useful to consider evidence not normally immediately associated with prehistoric conflicts, in order to assist in attempts to widen the base for archaeological research designs into the topic in the future.

The kinds of symptoms which may be identified include those which are commonly cited as evidence for warfare - weapons development and arms races, the design and employment of fortifications, aggregation of settlements for defence, the relocation of settlements to defended or defensible sites (and the concomitant abandonment of undefended or indefensible settlements), the emergence of formalised warrior castes, and the appearance of warrior or weapon-related iconography and symbolism.

Less often discussed are signals such as increased group identity confirmation and propaganda, the adoption of threat postures (such as the development of boundary defences), realignment of production and storage activities, and the abandonment of more remote or vulnerable areas in the build-up to aggressive action.

Osgood claims that "The archaeological evidence for war in the British Late Bronze Age (c 1200-700BC) is … strong. We have weaponry which seems to have been used in combat. There are hillforts … whose origins are now known to lie in this period. In addition, we have human skeletons with weapons injuries in spite of the fact that late Bronze Age human remains are scarce" (1999:8). Such evidence, for any prehistoric period, is certainly indicative of the possibility of warfare, but can it actually be said to be *strong*? Does it in any way *prove* that warfare had been taking place? In fact, examination of the evidence throws up more questions and doubts than answers. In order to assess the strength of the evidence, it is necessary to examine the counter-arguments and alternative interpretations that may be presented for it. From this starting point, it may be possible to identify the criteria required for confirming the strength of the evidence or improving its validity, and perhaps to begin to recognise the specific conditions that would be necessary to produce evidence that would truly 'stand up in court'.

Weaponry and Armour

Can the presence of arms and armour in the archaeological record take us any substantial way towards proving the existence of warfare in a society? Inter-personal weapons (as opposed to hunting weapons or tools) certainly must imply the possibility of conflict, the understanding of what armed fighting means and what its results could be, but there are many levels of conflict and of weapons use, not all of which, even with the most generous of definitions, can be characterised as war.

It is worth briefly considering the range of weaponry most usually found in prehistoric contexts, and its implications for evidence survival, recognition as a weapon as opposed to a tool, and for an understanding of both the society associated with the forms in general and that society's readiness for and forms of conflict in particular.

The simplest weapons are the 'found' forms - sticks and stones. These do not necessarily require any human modifications to become dangerous or even lethal. Studies of chimpanzees have noted their use of weapons of this nature not only for displays of fear and aggression, but for actual bodily harm and even killing of members of other troupes (for example, Goodall, 1990). Crush damage on bones might be expected as evidence of the use of such weapons. Split or sharpened sticks could produce sharp trauma to soft tissue, which would be unlikely to be represented on skeletal remains. Sticks used in this way are also highly unlikely to be found by archaeologists (with a few possible exceptions) (Coles et al, 1978:13), and while stones will certainly still survive, their identification as weapons would be very difficult. The use of found weapons might generally be expected to suggest relatively close range conflict, and, due to the spontaneous nature of the implements, might imply unplanned, emotionally or instinctively derived aggression. Ethnographic accounts of tribal conflict record the use of such weapons as a regular feature of endemic raiding and fighting, (e.g. Chagnon, 1968:171). We can expect these kinds of weapons to have been ubiquitous, but archaeologically they will remain almost invisible.

Modification of sticks can produce a range of other weapons; sharp objects can be embedded into the wood to form a mace, but the most usual alteration involves affixing a sharp point to the end (made of stone, bone or antler, or metal) and turning the stick into a spear, lance, javelin or arrow. With such weaponry, we are clearly dealing with hominids, because of the cognitive processes involved - the visualisation and realisation through a number of stages of an innovative product. The first development of these artefacts was almost certainly as a form of tool for hunting. There is a myriad of various forms of spear- or arrow-head in the archaeological record, and more rarely, but not unknown, there are examples of their shafts and of bows (Coles et al, 1978). Against a human foe, these weapons have a wide range of possibilities for close, medium and long range use.

When dealing with spear forms, there are a number of problems of identification, due to the wide variability of the evidence, and confusion of nomenclature. In addition, it is to be supposed that, when occasion demanded, a weapon designed for one form of use would readily have been employed in another. Definitions have been attempted for British and Irish spear forms [e.g. Greenwell & Brewis (1909) and many later classifications]. Some differentiation may be possible by examination of the socket forms of metal weapons and the weight and hafting arrangements of stone weapons. As Osgood points out, however, "..the smaller spearheads would have been just as deadly in thrusting mode" (2000:25).

Randsborg has analysed the spearheads from the ritual deposit in the bog at Hjortspring, Denmark and suggests that they represent a recognisable standardised set of equipment for a Pre-Roman Iron Age boatload of raiders. The material collected includes 65 iron-tipped 'lance' heads, 65 iron-tipped 'javelin' heads of which 31 had a broad head and 34 a narrow one, and 31 bone- or antler-tipped 'javelins'. Calculating a crew of 18 warriors and 4 commanders or steersmen per boat, Randsborg suggests that each man would have held one or two light throwing 'javelins' or 'lances' and one heavier, thrusting weapon for close-combat fighting, the whole find representing the armaments of perhaps three or four raiding ships, one of which was preserved in the bog. He also notes the presence of a number of flints, which he suggests were used as projectile weapons either to be thrown while the warriors were still on their boats (loss of such stones in the water being less costly than loss of manufactured weapons) or as the opening bombardment of a land engagement (Randsborg, 1995). This analysis of weapons tactics chimes well with Tacitus' description of the armaments of the Germans -

> *"Only a very few use swords or lances. The spears that they carry have short and narrow heads but are so sharp and easy to handle, that the same weapon serves at need for close or distant fighting. The horseman asks no more than his shield and spear, but the infantry also have javelins to shower, several per man, and can hurl them to a great distance"*
>
> (tr. 1948:105).

Randsborg (1995) considers the arrival of the trained cohort style of fighting force in Northern Europe a possibility from this time (third to second century BC), based on the Hjortspring evidence. A later hoard, from Ejsbøl Mose, southern Jutland, dated to the third century AD, has been analysed to demonstrate a force of over two hundred men:

> *"The combination of weapons at Ejsbøl suggests a hird or army with ranks and a command hierarchy, specialised arms, and mounted commanders. Its striking power is based on the strategic deployment of various weapons, infantry and horsemen: close-range weapons for hand to hand fighting combined with attack from a distance by arrow and spear"*
>
> (Hedeager, 1987: 132).

If this evidence is accurate as an interpretation of troop components, then the army represented is an heir of the warrior boat crews described by Randsborg, and by virtue of its organisation and specialisation, must clearly demonstrate the existence of warfare.

There is plenty of evidence for the use of these implements against humans in the palaeopathological record. Many assumptions have been made about their combat use but little tactical analysis has been undertaken for examples from Northern Europe (with the notable exception of Randsborg's study). Both throwing and thrusting weapons would have been equally efficient for hunting (and, indeed, there is plenty of evidence that they were); but apart from mass deposits such as those at Hjortspring, Krogsbølle and elsewhere, most finds are of limited numbers or single examples and thus are hard to identify as weapons of war. Some examples bear fine decoration, and may be seen as symbols of rank, wealth or status. This does not preclude their use in war, but it does not prove it either.

Many of the same remarks apply to archery equipment, and there are many examples of arrowheads found embedded in human bones or in close proximity in graves. The bow is, however, clearly a longer distance weapon. The use of archery in a formal battle is well known, but it is also a weapon of ambush, of secrecy where necessary. It also has some constraints. The bow is a skill weapon. Granted that to throw a spear accurately over any distance takes practice, it is still a much easier proposition than accuracy in archery. A good archer needs strength and technique as well as intention; the number of good archers in any community is likely to be less than the number of individuals able to wield a spear. Evidence for the use of bows in large scale conflict comes from a number of sites, such as Crickley Hill in Gloucestershire, where hundreds of arrow-heads were found clustered around the two entrances *(Figure 1)* and along the line of the palisade of the Neolithic enclosure (Dixon, 1989). At Carn Brea in Cornwall, *(Figure 2)* in the earlier third millennium BC, the defended settlement was destroyed by fire perhaps during an attack. Here, some 751 leaf-shaped flint arrowheads may also bear witness to an archery onslaught (Mercer, 1981).

Arrow heads are frequently found in graves without any clear relationship to the body of the deceased.

> *"If we pause to consider the circumstances of the burials in question, frequently disarticulated, often re-ordered, and very often disturbed by animals and later human activity, it seems perverse (and curiously archaeo-centric) to regard the very occasional, and often single, arrow as a 'votive deposit', as opposed to its arrival in the grave within the corpse, either as a cause of death or as a long-term legacy of some earlier conflict"*
>
> (Mercer, 1999:149-150).

Another form of artillery weapon is the sling, and deposits of slingstones have been found particularly in association with Iron Age hillforts. Maiden Castle has produced large amounts (Wheeler, 1943:XVI: Sharples, 1991:83) and at Danebury, great numbers were found (Cunliffe, 1993:48).

The sling as a weapon was used in the Classical world, and there is a depiction of close-ranked soldiers using an underarm action to discharge slings on Trajan's Column. In skilled hands, the sling was also a weapon of great accuracy, but

> *"the biblical story of David and Goliath, of a single stone killing a chosen target, has tended to mislead. Except at close-quarters, one single sling-stone is of very little use against one single individual. The effective use of missiles is to fire them from a distance, laying down a dense barrage over the general area of a static target. The aim need not be precise or accurate, and firing may therefore be more rapid"*
>
> (Avery, 1986:223).

Caesar's account of an attack on a hillfort in Gaul (tr.1951:II.3:56) involving a strategy of first clearing the ramparts of defenders by means of a hail of stones, followed by the firing of the gate is cited by Avery (ibid) as evidence of a familiar and well-used strategy. The large deposits of slingstones claimed at hillforts and elsewhere, if properly identified, certainly seem to swing the balance of evidence in favour of defence against massed attacks at the sites; whether these attacks were as part of a war, or occurred as occasional raids, is an unanswered question.

Axes of stone and metal are also frequent finds; their forms are, if anything, even more varied than those of spear- and arrow-heads. This is not surprising, as the axe was the favourite prehistoric tool in any number of functions, from tree clearance to butchery to boat-building. Any or all of these could also be used as a weapon and have been designated as 'Weapon-Tools' by Chapman, (1999: 108). Woodworking adzes were apparently used to commit the killings at Talheim, a late Linear Pottery settlement where at least 34 people were massacred and buried in a mass grave (Wahl & König, 1987). Some axe forms, however, seem to have developed specifically for use in combat; polished stone shaft-hole battleaxes have little obvious practical use, and the shafted halberds and some palstaves also seem to have been designed for conflict. How their use in warfare could be proved is highly problematic. Numerous studies have concentrated on the undoubted status attributes of axes, particularly the fine imported polished stone varieties of the later Neolithic; associated with status is the accumulation of wealth represented by their ownership. Chapman suggests that "the possibility that hoards of Neolithic polished stone axes represent not only communal wealth but also caches of weapons in readiness for raiding or warfare opens up new interpretational possibilities" (1999:110). Examples of polished axes that show absolutely no signs of wear or use, or are made of material far too brittle to be used in anger (such as alabaster) are common, as are finds of vastly over-sized or miniature examples. It is clear that the axe *form* had meanings beyond those of practicality and as such it appears in contemporary iconography (e.g. at Gavrinis, at Stonehenge, at Locmariaquer and elsewhere) as a symbol we still have to interpret successfully.

The greatest number of weapons hoards date from the Bronze Age. Possession of a metal weapon probably carried with it a certain cachet (demonstrated by the careful production of flint daggers in imitation of metal forms) and the mythology of blade weapons lives on to the present day. Blade weapons include knives, daggers, dirks, rapiers, swords and sabres. They are all close range weapons. Knives, daggers and dirks have functions as tools; the sword forms do not. Of all the weapons in the prehistoric armoury, it is the sword that demonstrates most clearly familiarity with inter-personal conflict. They are "the first weapons uniquely devised for warfare" (Bridgford, 1997a:6). Bridgford argues against the tendency to regard swords as largely ceremonial; her study of several hundred Irish Late Bronze Age swords (1993) has revealed edge damage and distortion of blades that may be indicative of actual combat use. Similar evidence comes from bronze swords studied in Hampshire and Hertfordshire (Colquhoun, 1979; Savage 1979) and from Denmark (Kristiansen, 1978). It is not always clear, however, that such damage could not have been inflicted as part of a ritual act. The Late Bronze Age sword, according to Bridgford, "must have been invented by societies already used to war, and aware of the limitations of other weapons" and this, combined with increasing evidence of Late Bronze Age origins for hillforts and other defended enclosures, "suggest that the Bronze Age must now be seen as a period in which warfare, and the threat of war, were deeply ingrained in daily life" (Bridgford, 1997a:6), but the evidence of the swords alone cannot be regarded as conclusive.

The Late Bronze Age sword is a culmination of developments, described in detail elsewhere (e.g. Allen et al, 1970; Burgess & Colquhoun, 1988; Parker-Brewis, 1923 etc.); the particular relevance of this information in the present study is for forms of fighting. It has been pointed out that the Earlier Bronze Age long blade weapon, the rapier, is a skill weapon, whilst the heavier later sword forms are melée weapons, needing little training for their use (cf. Bridgford, 1993:72-3); the remarkable Bronze Age yew sword from Grotsetter, Orkney (illustrated in Coles et al, 1978, fig. 3.1, p8), might be interpreted as a training weapon, although it could equally well have been a toy. It might therefore be inferred that the rapier was a duelling weapon, employed perhaps by chiefs or champions in small numbers, while the heavier slashing swords of the Late Bronze Age were more widely distributed among warriors in larger bands engaging in a less formal and more general battle style. It has also been suggested that some Late Bronze Age swords were designed for use by cavalry, adding a new component of mobility with increased shock force to warfare in the period. The evidence for this is discussed by Osgood (1998:14). Swords of the Iron Age are often associated with highly decorated scabbards; in Britain they have been found with high status burials and most commonly with ritual deposits in wet locations, particularly the River Thames and votive sites such as Llyn Cerrig Bach, Anglesey. However, iron survives less well than bronze in most depositional conditions and the surviving examples represent only a very small number of those that were in circulation, particularly, perhaps, more ordinary weapons of lower status warriors. This division of weapons into high status and more commonplace has been noted for the Bronze Age; "...in Scandinavia Period 2 - highly decorated swords found in complex grave groups show little signs of use, whilst a second group, which is known from simpler burials, reveal traces of combat damage and even of resharpening" (Bradley, 1990:80). Randsborg's Hjortspring study identified 11 single-edged swords, which he proposes represent the weaponry of 'commanders' or 'steersmen' in each of the three or four boat crews whose equipment may be contained within the bog (1995: 38-52).

It is in the development of the sword forms, and in the apparent variation between 'parade' and 'workaday' weapons that their actual use, as opposed to mere existence as display items, can be inferred. Whether or not this use was aggressive (as opposed to sporting) or in warfare rather than in more or less ceremonial duels, is still to be determined.

Armour finds - (helmets, shields, cuirasses, greaves etc.) are also concentrated in two main discovery contexts - graves and watery sites. Surviving examples of armour are predominantly rather special items, highly decorated and in some cases apparently unsuited for actual combat use. The thin metal comprising the Battersea and Witham shields, for example, does not seem appropriate to counter blows from iron weapons because of the damage that would be done to such expensive material. These pieces are usually designated as 'parade armour' and are seen as ostentatious forms of display by members of aristocratic élites related to wealth and status rather than to warfare. The differentiation of various forms of shields has been traced in the Bronze Age of Northern Europe and the evidence for their use summarised (Osgood, 1998). Damage on the thin bronze example from Long Wittenham, Oxfordshire, has been associated with perforation by a spearhead (Needham, 1979:115) and further examples of damage have been noted in shields from Germany (ibid:127). Needham (and Osgood, 1998) believes that bronze shields could have been used for both ceremonial and actual combat use, whereas Coles (1962, and Raftery, 1982:16) felt that the amount of decoration and the flimsiness of their metal consigned them to ritual or ceremonial use alone. Both Osgood and Thorpe (2000) note that weapons damage to shields may result from ritual 'killing' of the object as it is prepared for deposition rather than from conflict, although the apparent repair of earlier perforations on the Long Wittenham shield may indicate "that the shield was a veteran of several combat encounters" (Osgood, 1999: 9). Another possibility might be damage inflicted during weapons training. Warrior training in the techniques of warfare does not necessarily imply the actual *occurrence* of conflict; the presence of trained soldiers may even be regarded as a necessary component for the maintenance of a state of peace.

Organic examples of shields are indicated from Ireland

(Hedges et al, 1991), and from Germany. Coles (1962) generally considered that organic shields would be far more effective than metal ones, and easier to use in battle; it may be that organic shields were far more common, but of course, they are far less likely to survive into the archaeological record.

The same patterns can be seen with other forms of armour, particularly helmets - in some cases apparently quite unsuitable for actual wear in combat [such as the "unwieldy" and "top heavy" Viksø examples (Osgood, 1998:31)] but there are some surviving examples of organic helmets which may have been more common, wearable and effective, including those depicted on the statues of the 'Terracotta Army' of Ch'in Dynasty China. Glob (1973:80) describes an elaborately layered and stitched cap found with the Danish Muldbjerg chieftain grave which was "designed, like a helmet, to protect the head against cuts and blows in battle".

Presence of weaponry or armour in graves may seem to be a clear indication that the dead person was a warrior; but unfortunately, this may not be as simple as it seems. Weapons are also found in female and child graves, where they are generally accepted as indicators of status or wealth (e.g. Peebles & Kus, 1977; Whittle, 1996). Härke's study of Anglo-Saxon graves came to the conclusion that weapons in child graves related to inherited status and to rites of passage (1990) and miniature weapons found in child graves elsewhere have been interpreted as toys (Derevenski, 2000:7) - but the cruel truth is that the amorality of children makes them effective soldiers and they are frequently used as such in modern society, let alone in the past. Conversely, there has also been evidence that some weapons found in adult male graves were antiques, perhaps heirlooms, and thus more likely to be indicators of status, although this does not preclude their use in battle as well. Each case must be assessed individually, particularly in view of the overall limits of examples and dating evidence for each period and society - all that needs to be noted here is that the presence of weapons in graves cannot be regarded as a straightforward source of evidence for the occurrence or prevalence of warfare in any particular context.

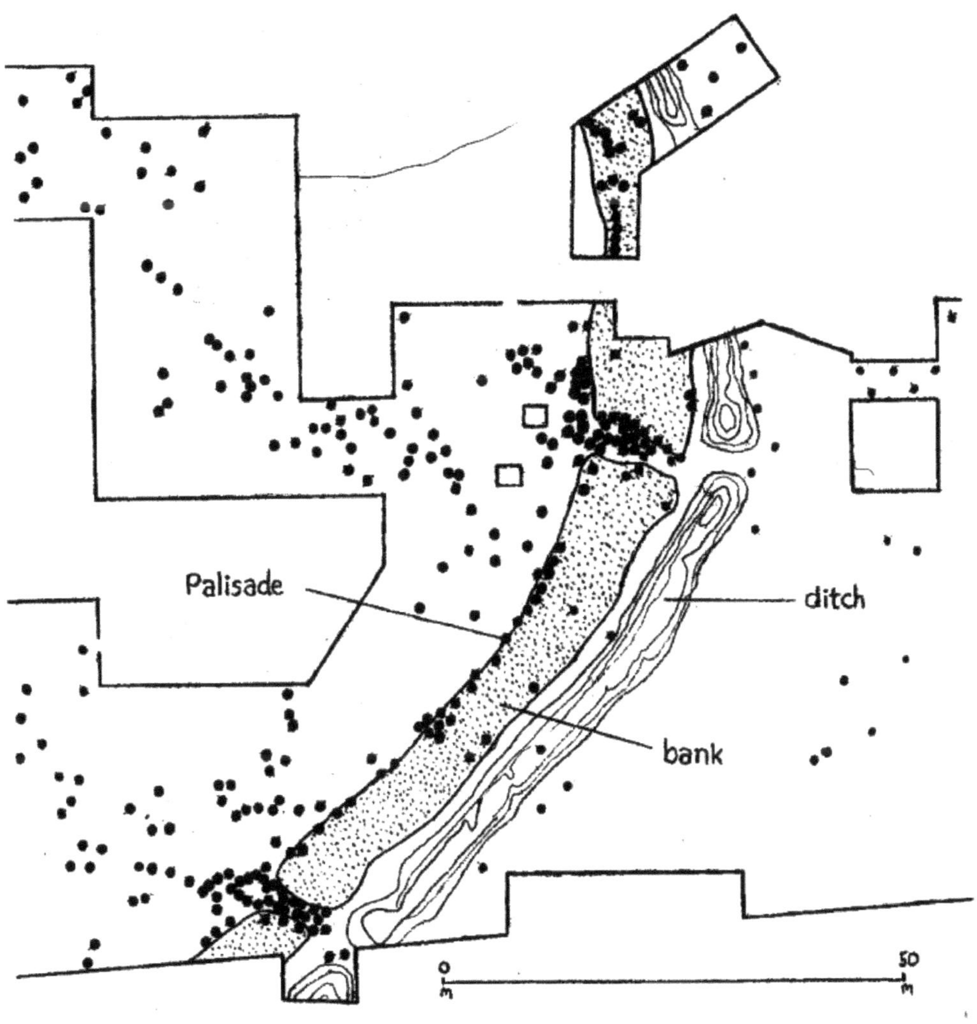

Figure 1: Plot of arrowheads recovered around the eastern entrance to Crickley Hill (from Hill & Wileman, 2002:19)

Figure 2: Carn Brae, Cornwall - a Neolithic site with evidence of attack by archery (from Hill & Wileman, 2002:20)

Depositions of weapons and armour in watery contexts, generally assumed to relate to forms of votive sacrifice, are not homogenous. In some cases, weapons are found with a range of other materials. In other cases, there is a generally more overtly military character to the deposits (such as Hjortspring). There are records (for example in Tacitus and Caesar) of the deposition of war booty in such places:

> *"After a victory they sacrifice the captured animals and collect the rest of the spoil in one spot. Among many of the tribes high piles of it can be seen on consecrated ground"*
>
> (Caesar, tr. 1951, I,1:34).

Randsborg (1995) supports such an interpretation in several Danish cases. More mixed deposits, comprising tools and jewellery as well as weapon items, such as those of La Tène, or Flag Fen, are less easy to understand. There is a further complexity of interpretation that should not be overlooked - recovery of artefacts from bog, river and lake locations rarely allows for unequivocal recording of relationships and stratigraphy. It may be possible that weapons from such hoards represent reuse of a previously sacred site for the deposition of war booty in one or more individual incidences interspersed with their normal pattern of use, or the later use of a site in which war booty from a particular event has been deposited as a form of memorial service. War memorials in modern society tend to acquire a form of sanctity long after the original conflict, and are often reused to commemorate the fallen of later conflicts in absentia - we may perhaps see some of the prehistoric contexts in this light. A further explanation sometimes offered is that hoards represent the concealment of valuable materials in times of trouble, but as Coombs (1975:68) asks, "why bury one's weapons in such a situation?"

There is a difference between having and knowing about weapons, and using them. They may have a ceremonial significance, or be used as a means of displaying particular levels of status or wealth, or as a badge of gender and maturity. Their presence need not of itself imply warfare; they may be used for ritual or sporting contests, or in the pursuance of crime, or be used only occasionally for sporadic raids or ambushes in small-scale fighting. Nevertheless, warfare generally requires weaponry and the employment of personal defences in the form of armour. In order to assess whether or not weapons and armour found in archaeological contexts can really be useful in the recognition of war, a number of questions require answers. First is the context of the find; where and in what circumstances these artefacts were deposited. This question is closely related to that of the motivations behind their disposal; why should people take weaponry (especially expensive, effective and highly elaborate examples) out of the realms of use and practical function?

Analysis of the artefacts themselves may tell us more about their suitability and effectiveness, and in some cases actual usage in combat may be established. Studies of the patterns of damage may help to illuminate something of the tactics and forms of fighting. Finds of weapons within bodies, or the forensic tracks of weapons damage to skeletal material show undeniably that use was made of them for person to person aggression. Evidence also exists for the use of organic weapons that rarely survive into the archaeological record but leave their traces on the bones of their victims. There are a number of other potential areas for study that could be significant, including temporal and geographical patterning of weapons use for the tracing of differing and developing tactical approaches and levels of aggressive intention.

Still, none of this evidence actually proves warfare, only its possibility. The artefactual evidence of weaponry alone can rarely prove the pursuance of contemporary war. It needs to be combined with the presence of other correlates.

Fortifications as evidence for warfare

Does the existence of a fortification necessarily prove that warfare was an endemic part of the society to which it belongs? True defences serve a variety of strategic and tactical purposes during periods of actual conflict. Keeley (1996: 56) has listed some of these fairly succinctly: fortifications act as a shield for defenders, noncombatant dependents, property and livestock; they make attackers more vulnerable to the weapons of the defenders; they prevent surprise attacks and infiltration; they tend to force attackers to concentrate on specific points like gates, mitigating any advantage in numbers attackers might have; and they provide a screen of manoeuvre hiding defenders' movements and strength. But "...defence against attack need not necessarily result in fortification, and the presence of physical barriers in or around settlements need not necessarily be due to defence" (Rowlands, 1972: 447). Protection of life, stock, and supplies may be a factor, but there are many levels of complexity beyond this that have been cited for defence-building. The decision to expend the labour and time, to say nothing of materials, on building defences must set the costs of the enterprise (which may also include inconvenient living conditions, limitations of mobility and social contact, and difficulties in conducting trade and other economic functions) against the perceived benefits; as has been pointed out, the building of defences has to be a cultural choice (Tringham, 1972: 470). Permanence of settlement and longer-term investment in the form of subsistence have been suggested as important features in the beginnings of fortification. In such societies, defences may protect against predators, or against incursions by non-sedentary groups. The building of defences may even help to end or limit aggressive behaviour.

The earliest recorded 'defences' so far noted are held by many to be those of Jericho dated to about 7500BC (Roper, 1975:309), while the evidence in Britain and other parts of Northern Europe for fortifications in the Early Neolithic rests on a number of excavated sites in southwest England and a few more in France and Belgium, where:

> *"The enclosure at Darion {6300-6200BC) included pasture and possibly a small grain field, which implies that the inhabitants were defending not only themselves but also the economic bases of their life. It may also imply they planned to withstand sieges of more than a few days. The character and complexity (e.g. baffle gates, multiple palisades, deep ditches, etc.) of Darion's defences seem clearly to have been intended to deter humans, implying that these enclosures were fortifications"*
>
> (Keeley & Cahen, 1989:168).

Keeley's interpretation of these sites can be challenged on a number of grounds; Keeley himself notes the 'incongruity' between the scale of the works and the number of inhabitants, and his colleague is rather more cautious in his assessment:

> *"A Darion, ces ouvrages relèvent d'une conception stratégique, sans doute élémentaire, mais qui dépasse l'intention d'empêcher la divagation des animaux domestiques et l'accès des prédateurs. Nous ne disposons cependant d'aucun indice quant à l'identité des assaillants éventuels"*
>
> (Cahen et al, 1987:67).

Keeley does suggest an enemy - "... the most likely assailants were the indigenous Late Mesolithic hunter-gatherers of the area" (Keeley & Cahen, 1989:174).

In Britain, the sites of Carn Brea, Hembury, Crickley Hill and Hambledon Hill are those usually referred to in any discussion of Neolithic fortification, although at Carn Brea,

> *"it may well be that the structure was deliberately constructed to present an imposing appearance, and was, in effect, a statement of the status of the community that built it."*
>
> (Mercer, 1981:67).

The presence of exotic goods and raw materials has been noted here and at Hembury in Devon.

Actual evidence for conflict appears to be proven at a late phase of the Neolithic settlement of Crickley Hill.

> *"The fate of this final enclosure was clearly shown by the thick spread of flint arrowheads, over 400 of which had choked the eastern entrance passageways and fanned out along the roadways into the interior. The enclosure has quite obviously been defended against archery attack, and it is highly likely that*

it was built with this intention, for the low palisade formed no more than a breastwork. The ditches were presumably designed to break up and slow down an assault, and the low bank, or rather platform, would then serve as a killing ground, at point-blank range, against aggressors clambering out of the ditch."

(Dixon, 1988:82).

A similar situation exists at the Stepleton enclosure by Hambledon Hill, (Mercer, 1988:104), where the conclusion is reached that "There can remain little doubt that we stand here in the presence of warfare on however limited a scale.."(ibid:105). This evidence has been challenged, both in the scope of possible warfare and in the interpretation of the specific events at these sites, specifically because these examples only occur, as far as is known, in the South West.

The idea that causewayed enclosures (those which lack the defensive developments noted at the sites discussed above) may have had a defensive role was suggested by Cunnington (1912) and Curwen (1930) but rejected by Piggott (1954) and Smith (1965); Drewett (1977) offered some defence of Curwen's viewpoint by criticising Smith's contention that the siting of the enclosures, the shallowness of some of the ditches, and the number of causeways necessarily negated the use of the interior as a refuge or arena for combat. He warns against the dangers of ethnocentricity in interpretation of Neolithic military strategy (ibid: 223-4) but tends to prefer an explanation that the original purpose at least of these sites was to contain exposure burials, and that they then developed individually with other functions. Other proffered explanations have been examined usefully by Evans (1988) and Vencl:

"That violence occurred at some sites is ..... some way from proving that causewayed enclosures were primarily intended to fulfill a defensive role.....there is little to indicate that the need for defence was a prime mover in the creation and early use of many causewayed enclosures. On the other hand, it may well be that the gathering of disparate groups at the enclosures, perhaps with livestock and exotic goods in their possession, proved the cause of raids or outbreaks of violence"

(Vencl, 1999:129).

The foundation for these other functions may be related to the observation (not universally accepted) that many causewayed enclosures were constructed on seasonal pasture, (Renfrew, 1973) and this writer has argued that each enclosure would be the focus for a chief and his followers, a centre for an economic hinterland with a base in stock-keeping and limited arable on the upland soil margins. Most writers reject the idea of organised chiefdoms at this early date, but some are willing to concede the development of economic territories centred around these sites.

The process of agglomeration of resources may be the reason for the kind of evidence found in the south west of Britain.

"The appearance of more complex earthworks and the construction of palisades might suggest that some sites acquired a more socially prestigious position with time, perhaps associated with greater restrictions on use and access. In these instances, some form of conflict may have ensued....during periods of local unrest arising from particular circumstances precisely because gatherings were taking place there"

(Oswald et al, 2001:129-130).

Other sites in other parts of the country also have evidence of palisades [such as Orsett, Essex (Hedges & Buckley, 1978)], and burned houses were found at Donegore Hill, NI, whilst concentrations of arrowheads were also found at the Billown enclosure, I of Man (Thorpe, 2000); Thorpe reminds us that organisation of causewayed sites for defence tended to occur late in their lives, a phenomenon he suggests may result from their preexisting ritual authority.

New defensive sites in the Later Neolithic in Britain are apparently unknown, the only two candidates (Mount Pleasant in Wessex dated to c2200BC and Meldon Bridge, Scotland, a palisaded enclosure with a date of c 3000BC) being regarded not as defensive but as ceremonial. For the British Chalcolithic and Early Bronze Age, no evidence of fortificationhas been found at all so far; however, substantial fortified sites are known for this period in the western Mediterranean with stone walls, bastions and redoubts; they have been seen as "large public works that suited the double purpose of material expressions of the territories' central places...and of monuments to an emerging leading power" (Oosterbeek, 1997:124) and "a political tool for social transformation" (ibid:127).

In the later Middle and Late Bronze Age, enclosed settlements with features interpreted as fortifications reappear in Britain, perhaps as a result of agricultural intensification and, possibly, population growth (Thomas, 1998:9), a viewpoint earlier expressed by Champion et al (1992:219)-

"The dwindling supply of available land for expansion made land a valuable item and created the conditions for more aggressive competition and raiding.... Warfare may only have been practised by a limited section of society, in particular by the elite, but it was assuming a greater role in the relationships between societies".

Once again, the military capability of these sites is questioned, as is any principal military purpose. Bradley (1978:119-120) suggests that Late Bronze Age hilltop enclosures lack effective defensive siting, and considers a possible ritual motivation in their construction. More recent discoveries of this type of site have produced further explanations but little consensus.

Osgood suggests that the imperative behind the construction of defended hilltop enclosures may be related to the recorded climatic changes of the Late Bronze Age.

> *"....The areas dominated by hill forts were probably those in which the greatest pressures had been felt, and sometimes where least could be achieved. The growth of defensive sites was possibly an aggressive posture accompanying competition for land, and perhaps raiding... One interesting possibility is that these sites were fortified food stores, central places in which an economic surplus could be collected and administered"*
>
> (Osgood, 1998:121).

There has been disagreement about the relationship between the appearance of these sites and the apparent contemporary proliferation of weaponry. Many have relatively poor dating evidence associated with their early phases, particularly those which developed later into full-scale Iron Age hillforts, and Bradley (1985:23) notes that despite the temporal coincidence between hilltop enclosures and weaponry, there is a spatial discontinuity between the two. In Yorkshire, some of these sites at least are recorded as having had earlier activity or meaning (Spratt, 1993: 128). Osgood conflates the idea of previous site status with the rise of chiefdoms or 'proto-tribal' organisation, the sites combining defence with displays of chiefly and/or corporate power and territorial marking (Osgood, 1998: 66).

The situation becomes no clearer as we proceed into the Iron Age proper. Those who believe the Iron Age to have been a warlike phase often cite environmental stress and land competition as major causal factors. According to Bradley, the lack of metal hoards in this period can be taken to indicate that what he describes as "surrogate warfare" in the form of competitive acts of metalwork deposition became irrelevant and was abandoned as a practice because actual conflict was taking place (Bradley, 1985:39-40). Cunliffe regards periodic bouts of warfare as an important function of early 'Celtic' society which was "effervescent.....essentially unstable and prone to conflict" (1993:53) and postulates a cycle of periods of peace followed by bouts of raiding leading to more general aggression which eventually exhausted the combatants who then returned to a state of lethargy, and claims that "the hillforts are the visible signs of just such a world" (ibid). Warfare was, he claims, necessary to establish status and had the secondary functions of boundary maintenance for the polity, and the acquisition of booty. These interpretations may rely too heavily on extrapolation from later sources, such as Dark Age literature; in the absence of any closer historical material (other than the equally unreliable Roman interpretations) however, the potential relevance of these views must at least be considered. In such an unstable situation as they suggest, the hillfort may be viewed as the "ultimate defensive weapon of European prehistory" (Avery, 1986:216).

Sharples offers an alternative view, believing that control of land was a vital element in the causation of conflict and the siting of hillforts, and that appropriation of the territory of others was a primary goal.

> *"The appearance of hillforts at these locations [with easy access to large areas of rich and diverse agricultural land] suggests that the land controlled was also actively fought over. The position allowed the establishment of a much larger community than was otherwise supportable from the agricultural resources, and hence gave this community an advantage in any conflict. Support for this argument comes from the knowledge that there are other adjacent areas where hillforts are* lacking"
>
> (Sharples, 1991:85).

While warning against acceptance of hillforts as a coherent entity, and suggesting that hillforts as a type of monument show little difference in terms of material culture and internal structures and activities from other forms of Iron Age settlement, Hill offers further explanation for this viewpoint.

> *"The need for communal defence would appear to be an obvious explanation for hill-forts ... reflecting warfare between large communities for control over land... The construction, defence and assault of hill-forts must have involved the whole community. Certainly hill-fort warfare (and construction) required the mobilisation of a significant proportion of the population. Such large masses for attack and defence, using essentially egalitarian weapons such as slings, etc., were far removed from the traditional picture of flamboyant 'Celtic' warfare and fit far better with farming communities organising to protect their collective resources in a period of apparent population pressures and expansion"*
>
> (Hill, 2001:107-8).

The arguments may not relate so well to hillforts in areas other than Wessex and Southern England, as noted by Hill (ibid) and Collis (1981:67).

> *"The hill-fort has been fundamental to viewing the Iron Age as a warlike age, even though endemic inter-communal warfare need not result in fortifications.. Assuming conflict between groups was common in the southern British Iron Age...,*

> *the question must be asked why only certain limited geographical areas used fortifications? ...warfare was just as endemic in [valley and lowland] communities, or at least they needed protection against their obviously warlike close neighbours. The absence of defended enclosures in lowland England points to a social explanation of hill-forts"*
>
> (Hill, 2001:110-111).

Citing the apparent long tradition of the use of monumental earthworks in Britain to delimit activity areas, to define community territory and to "mark significant social discontinuities in space", he suggests that hillfort ramparts should be regarded as a means of separating their interiors from "the cultural/social norms of the surrounding landscape" (ibid 112). In support of this view, he notes the apparently unmilitary tendency to orient hillfort entrances east or west, rather than in strategic relation to topography, the paucity of evidence for attack, and the argument that such evidence where it exists may be subject to alternative interpretation. Bowden & McOmish have suggested that evidence of the slighting of defences may be in some cases associated with ritual destruction by the builders (1987: 78-9) and that elsewhere the fact that interiors of some hillforts may be overlooked by observers from neighbouring slopes renders them undefendable and thus more likely to be intended for ceremonial purposes than for militarily effective defence (Bowden & McOmish, 1989:13). It is unclear just what kind of "ceremonial purposes" are being suggested, and it is this failure which undermines these explanations. Military explanations, whatever the tactical shortcomings of the sites, have, at least, the advantage of having a more identifiable function; "the theory explaining prehistoric fortifications as having primarily a symbolic function contradicts the need for self-defence that is in evidence from the Neolithic" (Vencl, 1999:70).

Sharples sees warfare of the later Iron Age as being more concerned with control over exchange networks and local alliances.

> *"These changes appear to correspond with a shift from warfare between local kinbased communities competing for access to local resources to warfare between the occupants of geographically defined territories ... [Warfare] would, of necessity, involve well organized incursions into distant and probably unfamiliar territories. Such expeditions would require careful planning, and they would also require the deployment of considerable resources"*
>
> (1991:85-86).

Associated with these changes are the evidence for much larger ceramic territories in some areas, a possible increase in economic specialisation and dependency on specialist activities. Whether or not the hillforts can then be seen as 'central places' for a polity has been the subject of much discussion; hillfort as élite centre is suggested by Cunliffe (1976, 1984 etc.) and Champion (1982:65).

The notion of élite centres was challenged by others including Stopford (1987) who saw Danebury as a communal defended storage place, and Hill (2001) who sees no evidence for increased status at these sites, while still others regard these monuments as status symbols for a wider communal identity (for example, Hingley, 1990).

The building of a defensive centre is often associated with the rise of complex societies; such centres feature among the hierarchy of settlement types and sizes suggested by Peebles & Kus (1977), who note the relationship between conflict and the rise of social complexity, and are also discussed in terms of their psychological and social impact -

> *"The choice to demarcate a settlement when there is no direct threat to the inhabited area or its contents may be the result of the attempt of one settlement to establish an illusion of impregnability and invincibility in the early stages of the formation of a political hierarchy of settlements... In addition, such physical barriers, which can be used for defence if necessary, serve as a clear demarcation of the administrative/political/religious/market centre of a settlement hierarchy, the urban centre"*
>
> (Tringham 1972:471).

The building of the large and highly visible ramparts can be seen as also having an internal function - "a way to discipline one's own workforce, keeping them producing goods for trade and under control" (Osgood, 1998:74). The actual military capabilities of hillforts are surprisingly seldom discussed in much of the literature. Avery has constructed an analysis of attack strategies based on hillfort morphology which would have involved specialised groups of assailants and formal organisation.

> *"This implies a highly developed, new form of warfare, stylised perhaps, but not in the heroic style of one-to-one combat. It implies an attacking army which can no longer be a mere cattle-raiding band or party of aristocratic warriors, but is highly organised, with special assault squads and in effect a separate commissariat arm, at least for supplies of ammunition"*
>
> (Avery, 1986:226).

This is suggested to be evidence for a society in which warfare was endemic and a social focus, dominated by a warrior aristocracy; and "outside the fort .. large armies were mobilised which could control and ravage the whole countryside" (ibid:228). We have Caesar's description of Gaulish attack strategies at hillforts, but beyond this there is little military analysis. Clearly, the problem of water provision in British hillforts would suggest that lengthy sieges were never part of the expectations of hillfort builders, and the lengths of the ramparts would indicate that defence against mass armies was not envisaged. The design and sophistication of entrance hornworks nevertheless seems to suggest a very real appreciation of and concern with military architectural planning. They form a confusing approach, which can be defended at a number of points; they have dead ends and areas that seemed to be designed as killing grounds; they form funnels that could congest attackers and render their numbers ineffective; they create routes that seem to be deliberately designed to force attackers to approach with their unshielded right sides exposed *(Figure 3)*. If these sites are ritual, or ceremonial, or intended simply as political statements, why go to so much bother? Why do so many sites show redesign of the ramparts and ditches in ways that seem to reflect changes in military strategy, such as the removal of berms, and the cutting of wide Fecamp-style ditches *(Figure 4)*? These questions remain unanswered and often ignored as evidence for real warfare. Colllis (1981:75) felt able to state that:

> *"I have never liked the equation 'hill-forts = conflict'. If we were to make a simple numerical count of surviving weapons in Britain, the iron age would come out as a peaceful period, especially in comparison with the late bronze age. The hill-fort may have been a successful solution to keeping the peace."*

Sharples also considers that hillforts cannot be accepted as clear evidence for warfare but that ".. the bulk of the evidence for warfare in the archaeological record is created as a deterrent, or to symbolise the nature of the conflict rather than actually practice the act" (1991:88), functions that involved ritualised display and threatening behaviour rather than real violence.

The problems of trying to equate the building of fortifications with the prevalence of warfare can therefore be seen as complex, and closely related to interpretations of the societies that constructed the monuments, their norms, values and complexity. Vencl (1999) also points out that much of the evidence is unspecific and has been subject to change and erosion in later periods. A

> *"substantial complication affecting the archaeological interpretation of defended sites] is the additional functions (economic, social, ritual, etc.) that a fortified site acquires through being a place of safety. This feature may even submerge its primary military purpose because of the different archaeological visibility of particular activities"*
>
> (Vencl, 1999:69).

He earlier specifies that

Figure 3: Defensive 'hornworks' at Danebury hillfort, Hampshire (source: author's photo)

Figure 4: Wide 'Fécamp-style' ditch at Danebury hillfort (source: author's photo)

> *"...features indicating that a structure could have a military purpose can be summarised as follows: the presence and differentiation of various types of entrances (gates) restricting or controlling approach; the presence of traces of fighting (weapons, destruction, fire); multiplication of defence-works, especially at spots with easier access, revealing a concept of in-depth defence; proliferation of walls by the insertion of additional elements such as bastions and flanking towers; and the choice of inaccessible locations uncomfortable for daily living but possessing good views that offered greater safety and easier defence. No one of the indicators, if present, is in itself conclusive evidence of a military structure, but neither does their absence mean the contrary"*
>
> (ibid:68).

In view of the discussion above, it is evident that the presence of a fortification (even one that fulfills all of Vencl's conditions) cannot be regarded as evidence of warfare in itself. Even where weapons and fortifications are found together, or there is evidence of destruction apparently due to conflict, the interpretation can be restricted to a local or site- and incident-specific phenomenon.

Skeletal Trauma

Skeletal trauma is often held to be the most secure evidence for interpersonal conflict available to archaeology.

> *"Although skeletons exhibiting weapons injuries are not necessarily proof of warfare (other explanations include executions, murders, immediate post-mortem wounds and even ritual killings), it seems probable that combat is the main explanation."*
>
> (Osgood et al, 2000:19).

The clearest evidence for conflict is generally considered to come from two main groups of material - skeletons with embedded weapon parts or mutilations, and mass graves. Unfortunately, the palaeopathological evidence alone can only rarely prove that the deaths under investigation were the result of group conflict. The biggest problem lies in our ability to contextualise the finds.

There is a large number of instances of bodies, particularly from the Early Neolithic of Northern Europe, but also found commonly in North America, which have produced evidence of embedded projectile points within bony material. Examples have been listed in a number of articles - a few will suffice in this discussion. Mercer (1999) points to what he considers to be a large amount of Neolithic evidence indicating archery as a major form of attack, noting, among others, examples from Cat's Water (Fengate) and Ascot-under-Wychwood and also notes the many instances of arrowheads being found in close association with skeletal material (for instance at West Kennet long barrow), suggesting that these may represent soft tissue trauma as a cause of death; Osgood (2000, 1998) lists Bronze Age examples of bronze spearheads embedded in victims from Tormarton (Glos.), Dorchester-on-Thames (Oxon.), Aveyron, Over Vindinge, and La Grotte de Pas de Joulie Trèves (Gard), and arrowheads at Klings, Barrow Hills (Oxon.), Cartuja de las Fuentes (Ebro Basin); Venta del Griso (Teruel); Hipogeo de Longar (Navarra), and Hogeloon. At San Juan Ante Portam Latinam (Alava)) up to 336 individuals were interred in the cemetery probably all at the same time, including eight identifiable skeletal injuries caused by arrows, dated to the 4th mill. BC (Osgood et al, 2000). According to Ferguson (1997:323), "the presence of many projectile wounds virtually confirms war".

That this need not be the case is argued in many accounts. Chapman (1999) regards the evidence from central and eastern Europe to be scarce and poor, as does Roper (1975) for the Near East, although this may be explained in a regional sense rather than negating the explanation generally.

Thorpe (2000) speculates on the likelihood of criminal justice, accident or ritual as causation in the death of the recently investigated Stonehenge individual, and in his discussion of trauma represented in bodies from Iron Age cemeteries in East Yorkshire including Garton Slack, Burton Fleming, Wetwang Slack, and Danes Graves, Dent considered it could have been due to civil penalties or violent attacks in peacetime (Dent, 1983:12).

Mass graves seem to offer better evidence for warfare, although once again the context is vital. "A communal burial alone does not constitute conclusive evidence for a violent cause of death. Other factors such as a serious infectious disease or ritual depositions may also explain why several persons died within a short span of time" (Smits & Maat, 1993:24-5). Among the better-known mass graves is at Talheim in Germany, where at least 34 people (nine men, seven women and 16 children and adolescents) appear to have been massacred. 18 of the skulls had indications of fatal injury, mostly damage to skulls by polished stone axes but also blunt force trauma and arrow wounds (Makkay, 2000). The arrow wounds would seem to indicate that the attack took place in daylight, and the absence of defence injuries suggest that the attackers met with little resistance. Many victims sustained wounds from behind.

Further examples come from Schletz, in Austria, and Herxheim, Rheinland-Pfalz (Vencl, 1999). At a Middle Bronze Age site at Velim, Czech Republic a series of large pits seem to contain the bodies of hundreds of people, some bodies complete, some disarticulated, scattered or incomplete; some pits contained groups of skulls, and a number of bones had evidence of cut marks. Men, women and children were represented here; Harding (1999) considers that this site must be considered as evidence of warfare.

One of the best studied mass graves is the 14th century AD Crow Creek site in the USA. About 486 individuals, thought to represent some 50% of the village population, were found in the remains of the fortification ditch; the palaeopathological evidence includes numerous examples of mutilation - nearly 90% of the skulls, including those of the children, showed evidence of scalping; it is also suggested that teeth and tongues may have been removed and noses cut off. Decapitation was fairly common and hands were often removed; cut marks were found on joints suggesting dismemberment. Many of the dead (around 40%) had suffered blows to the skull and others died within burned houses. Canids had had time to gnaw bones before burial. It is believed that the attack took place in the late autumn or early winter, based on analysis of the process of decomposition. The moment for the attack seems to have been chosen while the village's defences were undergoing refurbishment:

> *"..six of the post holes of the inner ditch contained human skull fragments. This suggests that the inner ditch holes may have been open at the time of the massacre, the posts having been pulled, perhaps to place them along the newer, outer ditch. The outer ditch may well have been under construction at the time of the massacre to protect an expanding village and to replace an old ditch that had lost its effectiveness because filled with village debris"*
>
> (Zimmerman, 1997:87).

The problems of establishing the facts of the case, even in such apparently clear cases of massacre as those recorded in recent history, are largely due to the casual disposition of bodies, the length of time they remain exposed before burial, and the degree of mutilation and destruction incurred by the bodies at or near the time of death. At Kibuye Church, in Rwanda, bodies were buried several days after death and exhumed by war graves investigators six months later. It was found that the remains of children and especially infants were so badly decomposed that their bodies were, in some cases, unrecognisable as such (Schmitt, 1998).

At El Mozote, El Salvador, the remains of 143 individuals including 136 children and adolescents and seven adults were disinterred

> *"... uncertainty regarding the number of skeletons is a resection of the extensive perimortem skeletal injuries, postmortem skeletal damage and associated commingling. Many young infants may have been entirely cremated; other children may not have been counted because of extensive fragmentation of body parts"*
>
> (UN Truth Commission, 1992).

Minimal postmortem insect activity and scavenging of animals proved that the deaths were roughly contemporaneous with each other. The massacre took place on 12th December 1981, and the exhumations were undertaken between 13th-17th November 1992. Similar difficulties have been reported from other such sites, including the dry well containing a minimum of 162 individuals (67 under 12 years old) at Dos Erres, El Petén, Guatemala, exhumed in 1994 (the massacre of this village having been carried out in December 1982) (van Lier, accessed 2002).

In the case of Crow Creek, it has been pointed out that:

> *"A count based solely on unambiguous morphological criteria surely omits some fatalities from violence. It is unlikely that all fatal injuries would have affected one or more bones. This problem is compounded by the fact that it was not always possible to determine the origin of bone damage....In addition, it cannot be assumed that all victims were recovered for burial in the cemetery. The disarticulated and carnivore-damaged skeletons show that their deaths often occurred in irregularly frequented places where remains would have been difficult to find."*
>
> (Milner et al, 1991:588).

Further problems are created by the imposition of different minimal counting systems; missing crania, for example, may represent ritual decapitation, accident, erosion loss, and, in the case of Crow Creek, the activities of burrowing groundhogs (Milner et al, 1991).

It should therefore, perhaps, not surprise us that there is sometimes reluctance to accept prehistoric evidence from mass graves as being proof of warfare. Cause of death, number of individuals surviving to be identified, postmortem alteration and damage, and uncertainty about contemporaneity of deposition all combine to detract from the certainty of warfare as the causal factor. Other interpretations of mass graves as *single* events (as opposed to the belief that a sequential pattern of deposition has been misinterpreted) have included epidemics, volcanic eruptions or other natural disasters, shipwrecks, and clearing of older burial grounds to make way for new bodies. In analysing the Early/Middle Bronze Age multiple grave at Wassenaar, in the Netherlands, Louwe Kooijmanns (1993) made a point of trying to exclude such alternative explanations in order to achieve a reliable analysis.

The late Neolithic mass grave at Esztetrgályhorváti in Hungary (*Figure 5)* was considered by the excavator to represent a cultic site, based on burning considered to be part of a sacrificial ceremony, vessels and a grindstone associated with ritual deposited in a superficial layer of the pit (Makkay, 2000:19-20). The excavator further cited parallel evidence from Friebritz (Lower Austria) where eight bodies were found near the centre of a ritual enclosure, with two more individuals, who had been killed by arrows, possibly as a punishment or for ritual purposes, and from Ragona, Slovakia where a pit containing six bodies displayed similar characteristics to those of Esztetrgályhorváti. Makkay disagrees - "there emerges the possibility that the male persons .. were victims of a war-like fighting and their bodies were buried in a conciliatory ceremony towards the souls of the killed enemies/prisoners as part of a postwar ritual" (ibid:64).

Post-war ritual may also be the cause of the curious Late Mesolithic deposits found at Ofnet in Bavaria - two pits containing carefully placed skulls of men, women and children. Frayer (1997: 183-4) calls this "the first clear evidence for a mass murder in prehistory". The skulls (about 31 in one pit, 6 in the other) were packed "like eggs in a basket", most facing west, stained with red ochre and in the cases of the females accompanied by pierced red deer teeth and shells. Most of the skulls show evidence of blunt trauma, deemed not to be due to postmortem factors because of shape and position on the skull. Care was also taken to eliminate marks caused by cleaning, measuring and reconstruction in the past. Clear marks on many skulls also show that victims were decapitated by first having their throats cut, then having their cervical columns cut through above the first thoracic (ibid:205). Ofnet has a high proportion of children compared to other sites (59.5% under age 15). Comparable figures for Talheim and Crow Creek are, respectively, 44.1% and 46.3% (ibid:207). Ofnet also has an extremely high proportion of women (66.7%) compared to Talheim (43.7%) and Crow Creek (45.3%). It seems clear that the skulls could have been deposited in a single incident. It would seem odd, in view of contemporary burial practice, that the bodies were not recovered along with the skulls if these were buried by relatives or survivors. It is therefore possible that the skulls were placed in their pits by their killers, perhaps as part of a cleansing or propitiatory ritual (as suggested by Thorpe, 2000).

Studies in the United States have identified several situations which appear to demonstrate an endemic pattern of raiding. The Onoeta cemetery at Norris Farms #36 was in use for a relatively short time; in the low mound were found the remains of 264 individuals, 16% of whom showed signs of violent death.

> *"Trauma lacking evidence of healing, cuts on bones from body mutilation, and marks from carnivore gnawing were present on these skeletons, which usually exhibited more than one kind of damage"*
>
> (Milner, 1995:225).

A number of individuals were mutilated - scalpings, decapitations and dismemberments were noted. These traumatic deaths accounted for 16% of the cemetery population, and were nearly all adults (34% of adult population). Some people were buried individually, but unusually for this culture, some multiple burials were also found:

> *"Multiple burials often consisted of bodies that must have been in a similar state of disintegration when interred, judging from the bones present and their disposition in the graves. These burials indicate that single-sex groups, probably work parties, were the targets of some attacks"*
>
> (Milner, 1995:227).

An interesting observation was made about general health - it was noted that among these victims of violence, there was a high incidence of debilitating conditions pre-existing, perhaps making these people less able to fight or flee. The pattern of burials suggests that there had been many different encounters, each resulting in a few deaths.

> *"The pattern of violence at Norris Farms is consistent with chronic intergroup warfare......"*
>
> (ibid:229).

This pattern seems to be repeated in other parts of the region, accompanied by palisaded or defensibly sited villages.

In Alabama, a Mississippian Period cemetery (AD 1200-1500) on the Middle Tennessee River produced 102 burials including four mass burials each containing 5-6 people; there was evidence of scalping on the remains of five men and two women. Again, single interments are the norm for this culture, so the mass burials stand out. One grave contained only women and children, and While males predominated in the others, at least one woman was always present (either as non-combatant victims or, indeed, as members of war parties - in historic times, one of the last 21 warriors in Geronimo's band of Apaches was a woman, Lozen). Some 21% of the total population of the cemetery had suffered violent death. "The recurring nature of these attacks suggests that warfare was pervasive in this region of prehistoric Alabama" (Bridges, 1996:73).

The bodies from Crow Creek also showed signs of poor health including all the signs of repeated, prolonged malnutrition - Harris lines in the long bones were common, and among the skulls studied, 18 had orbital cribra, 4 had orbital cribra associated with other skull lesions and 6 had porotic hyperostosis. Zimmerman associates this with climatic changes, especially drought, which has been recorded for this period.

> *"The total picture which has emerged ... is one of long-standing chronic iron deficiency anaemia,*

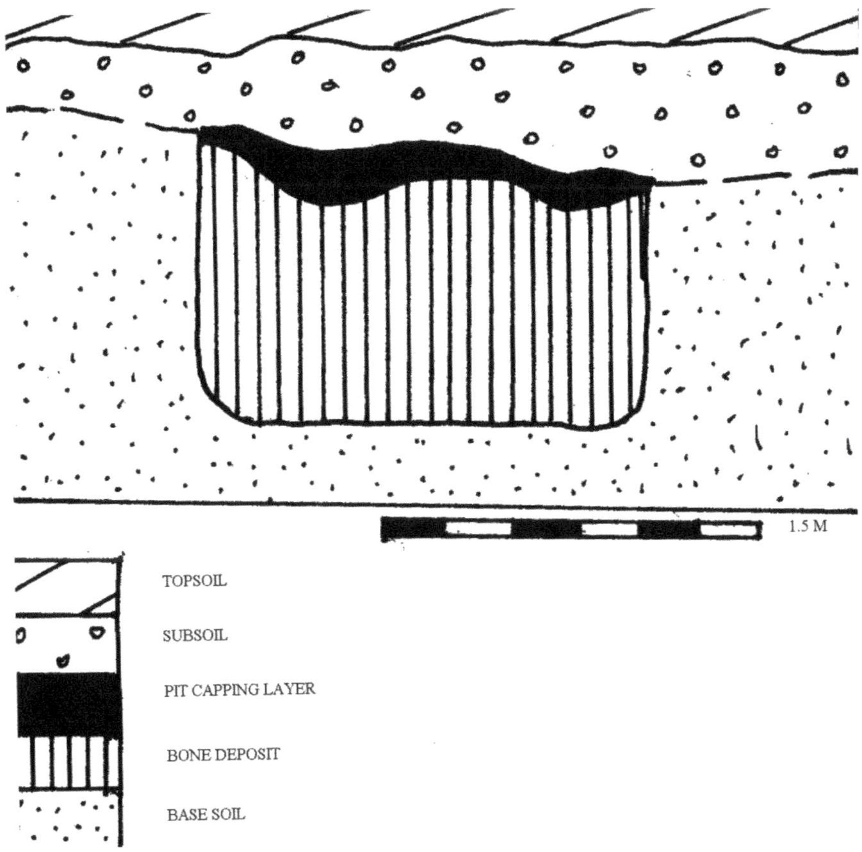

Figure 5: Sketch of mass grave at Esztergályhorváti (after Makkay, 2000:79)

famine, profound malnutrition, and scurvy. These conditions do not appear to have been short-term or temporary."

(Zimmerman, 1997:89-90).

Continual warfare, it is suggested, both causes and results in dietary stress in a small population; fear of attack may reduce opportunities for food gathering, raids may deplete herds and stockpiles, and part of the work force may have to be diverted to protection (Milner et al, 1991:595) and so

"Warfare had a demonstrable impact on the survival potential of individual communities, and some groups were utterly devastated by such conflicts.... the osteological evidence is compatible with the portrayal of warfare between small-scale societies that is presented in much of the ethnographic literature"

(ibid:236).

Milner (1995) has produced a list of criteria that are used in his analysis to confirm this pattern of constant ambush and small-scale conflict: the pattern of violence is of the hit-and-run kind; more adults die than juveniles; both men and women were killed but only a few victims at a time; wound locations suggest people were ambushed by opponents using commonplace weapon-tools; the dead are often buried in village cemeteries and there is evidence of mutilations; finally, some people occasionally survived attacks.

"From an archaeological perspective, mutilated skeletons provide compelling evidence of prehistoric war, since few societies would mutilate their own dead" (Keeley, 1996:100). Keeley suggests reasons for skeletal mutilation and removal of body parts -

"In some [societies], they merely provided a tangible numerical measure of a warrior's prowess. In others, they possessed magic powers that strengthened their possessor or transferred the victim's spirit to the victor's benefit. They might be necessary paraphernalia for rituals honouring deities, initiating youths, or cleansing their taker of the spiritual pollution of homicide. These items might degrade the victim, injure his afterlife, or enrage his survivors..."

(ibid:101)

and also notes the occurrence of 'overkill' to express hatred, or to enrage survivors.

For comparison, however, it might be educational to consider some more modern evidence, this time from Malawi. A *Times* correspondent, reporting on the current maize shortage and famine, recorded attacks on thieves stealing unripe cobs to feed their families. He quotes Dr Steve Mannion, a British surgeon working in the area:

'"People who are caught stealing maize are frequently subjected to mob justice, which typically takes the form of multiple, severe lacerations inflicted with panga knives," Dr. Mannion said. .... "The cases are all male, with an average age of 40, 90 per cent of whom are unemployed. The commonest sites of wounding are the forearms, kneecaps, ankles and skull. Two patients have suffered traumatic amputation." and "The people we were seeing were not normally thieves or robbers. They were from a totally different section of the population. They were invariably dressed in rags and have been driven to stealing green maize from their neighbour's fields in a desperate bid to feed their starving families."

(Dynes, 15th July, 2002).

The injuries described in this report could, in a prehistoric context, be interpreted as those of victims of ritual mutilation; the actual motives are much simpler - a person without hands or feet cannot continue to steal maize. If they die, they will end up in village cemeteries, perhaps in single graves or perhaps, given the general weakness of the population and the number of incidents, in communal graves within those cemeteries. What will archaeology make of their remains several hundred years from now?

Amputation may also have occurred as a means of counting the dead [cf. Roberts & Manchester (1995:89-90) on early Egypt], as a medical treatment or as a judicial punishment. Scalping may also represent a medical procedure or a ritual behaviour - Herodotus describes the preservation of scalps as amulets. Other grey areas include self-inflicted injuries, obliteration of evidence by bony regrowth, genetic conditions such as biparietal thinning causing foramina or cranial dysphrasia, damage caused by cysts and tumours, and scarification and cauterising techniques intended to relieve the symptoms of epilepsy, migraine or madness (Wakely, 1997).

The need to contextualise these finds has led firstly to attempts to define injuries due to intentional attack morphologically, and secondly to a call for more general assessment of the health and environment of the members of a cemetery or grave population. Noted above are the problems associated with the identification of the number of victims of mass graves; nevertheless, some agree that "Unhealed trauma, mutilation, carnivore damage, body disposition, and differential skeletal articulation permit the reconstruction of the prevailing pattern of violent death" (Milner et al, 1991:589)

On the other hand, Ferguson (1997:329) maintains that sports, spousal abuse, non-lethal conflict resolution, ancestor cults, secondary burial, and vagaries of presentation can contribute to misinterpretation; even apparent haphazard burial may be suspect - "casual disposal may indicate that the dead were from a different group" (1997:323), not necessarily involved in aggression. Assumptions based on pure numbers of cases can be inappropriate, reflecting only changes in mortuary practice in general - when inhumation is the norm in a society, evidence of trauma is far more common than when cremation is usual. This does not mean that periods of cremation burial are necessarily more peaceful or vice versa.

In a study of six male skeletons from an Anglo-Saxon cemetery in Eccles, Kent, Wenham (1989:127) categorises the diagnostic criteria needed to positively identify edge-weapon injury:

> *"1. Linearity, without large irregularities in the line of injury.*
> *2. An edge to the injury which was well-defined and clean.*
> *3. A cut bone surface which was flat and smooth and in some cases polished.*
> *4. The presence of parallel scratch marks on some cut bone surfaces".*

Further analysis of the length of cut marks and their direction, together with experimental use of replica weapons and fresh bone surfaces, was taken to indicate that swords were the weapons used to create the observed damage.

Similar analyses of blunt trauma have been carried out; there are many examples of fractures and depressions in cranial material in particular, although not all of these need have been fatal.

> *"Often the most dramatic injuries are to the skull as, certainly in interpersonalintergroup violence, the head is often the main target for blows.... it is the head which is usually the target for violence as the brain and skull are the most vulnerable areas of the body when damaged"*
>
> (Roberts & Manchester, 1995:79).

Whilst this is probably true, Walker (1989:320) has noted the common occurrence of self-inflicted injury to the forehead in ethnological accounts, associated with healing or expressions of grief.

Studies of blunt force trauma in a sample of 744 skulls in the Channel Island area of Baja California suggests that these injuries "seem to be the result of fighting intended to injure but not kill" (ibid:319). Ritualised duels involving two protagonists standing face to face and exchanging blows in turn are not unknown in the ethnographic record and relate, perhaps, to the historical Plains Indians custom of 'counting coup' - landing a non-lethal blow upon an opponent satisfied honour and accorded recognition of prowess.

A study of material from a sample of 80 Mesolithic sites by Constandse-Westermann & Newell (1984) found just five clear cases of interpersonal violence which involved projectile points embedded in bone, and concluded that in most cases, reported trauma was caused by accidental injuries; for the Neolithic, there was even less evidence, which they tentatively suggest may be due to people leading a more sedentary existence.

Conversely, in a consideration of trauma among palaeolithic individuals, Wells (1964) concludes that interpersonal violence was typical. "Antelope humeri have also been recovered in association with [australopithecine] skulls and the large articular condyles of these limb bones fit neatly into the double-valleyed depressions in the cranial fragments. There is no doubt that the heavy ungulate humeri were admirably suited for use as clubs and that a million years ago the precursors of sapient man were already nimble at battering their fellows..." (ibid:45). However dubious this assertion, Brothwell (1999), using similar evidence, has also come to the conclusion that trauma observed in some Neanderthal skeletons is unlikely to have been accidental.

Cannibalism is also often taken to represent the occurrence of warfare. However, eating of bodies does not necessarily imply violent death, any more than human sacrifice need represent coercion. As Ferguson (1997) points out, people can volunteer for these treatments, in accordance with their perception of the greater good. The criteria for the recognition of cannibalism have been refined in recent years and now include brain exposure, facial mutilation, burnt bone, dismemberment, pattern of missing elements, greenstick fracturing of long bone shafts to expose marrow, cut marks, bone breakage, anvil or hammerstone abrasions, many missing vertebrae, fragment end-polishing derived from cooking in a coarse ceramic pot and the presence of human myoglobin in faeces (Chatters, 1989). "Those archaeological cases that do qualify all appear to involve aggressive cannibalism, inflicted on defeated outsiders..." (Parker Pearson, 1999:53), but it is pointed out that this need not be surprising because other forms of this behaviour are less likely to result in so much damage (including behaviour aimed at symbolic communion, acquisition of prowess or knowledge, or to release the soul of the departed). Pressure may also exist to disregard apparent evidence of cannibalism because of modern indigenous sensibilities and political attitudes.

There has been a tendency for a somewhat blinkered approach to the study of ancient trauma - "Detailed studies are only made about past violence when it is believed that the cause of death is known. In practice this usually means that either a wound is identified which can be attributed to a particular weapon or when death was caused by a rare disease" (Arnold, 1995:71). The importance of a biocultural approach has been stressed by Brickley (2000:142) - "The environment which existed during an individual's lifetime is crucial to interpretations made about data derived from human skeletal remains" who goes on to suggest the usefulness of studies of disease, diet (using carbon and nitrogen isotopes), DNA to establish relationships of groups to one another, and stress indicators manifested in skeletal and dental pathology. The usefulness of more general population surveys can include the recognition of widespread change in diet and levels of general health caused by population dislocation, change or contractions in the subsistence base, and population movements. More specific study may be able to recognise those who died as a direct result of conflict but from illnesses that do not involve skeletal damage. Infection would probably be a major cause of death even from superficial wounds. Roberts & Manchester (1995) point out that only gross injuries are normally recognised, but that for most people, such injuries are rare. Infections may have been much more commonly implicated in mortality rates, but archaeologically, these are very difficult to recognize.

Other infections are likely to have occurred as a result of military behaviour, especially where campaigning in open territory was undertaken. People crowded together in insanitary conditions (whether soldiers or refugees) are particularly prone to tuberculosis, various treponemal diseases, parasitic and ectoparasitic infection, and endemic and venereal syphilis (a particular consequence of rapine and indiscriminate contact with camp followers). One of the most prevalent war-related infections is typhus:

> *"The infectious disease of typhus is a serious, sometimes fatal, sequel to life in insanitary conditions. Also known as camp fever, typhus was well known to campaigning troops in recent wars. No doubt it was common too in earlier times, particularly among the besieged and the besiegers in prolonged medieval warfare. But there is no skeletal or mummified evidence of this acute infectious disease."(ibid:162).*

Evidence for infectious diseases does not remain in skeletal material in a way that is currently recoverable, but modern applications of DNA analysis may eventually provide some form of identification of some infectious conditions.

Possible stress markers for displaced or endangered population groups include spontaneous abortion rates, infertility, dietary deficiency conditions such as scurvy and rickets, dental enamel hypoplasia, cribra orbitalia, Harris lines, osteoporosis in young people, and overall decrease in size and weight. Some of these conditions are visible on the skeleton; others may be identified through the analysis of trace elements and stable isotopes. Isotopic and trace mineral analysis (particularly of heavy metals) may also be used to demonstrate population movements - particular concentrations of these elements are sometimes restricted to quite specific locales. If we were to find communities where the children's pattern of these features is markedly different from that of their parents, we could suggest in some cases the kind of demographic movement that may result from warfare.

That warfare can create the conditions for general population distress is, of course, well known. Referring to ethnographic evidence from the Mae Enga of New Guinea, Meggitt (1977:12) has observed:

> *"We simply do not know how many infants and old people succumb to pneumonia in these flights, how many refugees are drowned when trying to cross boulder-strewn torrents, how many already sick and weak people die because food supplies are interrupted. These less obvious costs of war, I believe, accumulate significantly through time ".*

Milner (1995:221) has said "Despite the rather obvious connection between skeletons and combat casualties, osteological analyses have made few contributions to the study of conflict among prehistoric peoples..". Where confirmatory evidence can be added to that of the bones, it seems reasonable to assume conflict, as at the Chalcolithic site of Valencina de la Concepción, (Sevilla) where two skeletons were thrown without ceremony into a ditch, largely disarticulated, one without head, the other a limb. "The seemingly violent nature of their death, the careless method of disposal, evidence of burning from surrounding areas and the possibility that the ditch formed part of the defence of the settlement perhaps all imply a violent attack on the site." (Osgood et al, 2000:46). Whether one can proceed to more generalised statements is less certain. The same authors go on to provide an overview of Bronze Age warfare that takes the rather scarce evidence a rather long way, into the depths of symbolic social structures. These claims may well be true - but they lack confirmation. Roberts & Manchester (1995) stress the limitations to palaeopathology; the recovered dead are not necessarily representative of the living population. We are working with a very small sample, we should assume differential disposition of gender, age, caste, status, ethnic origin, etc., and be aware that preservation is often poor, especially of non-adults; in most cases the cause of death almost impossible to establish. The nature of the evidence is patchy, not only in terms of the finds, but of the standards of reporting and analysis, especially where older reports are concerned (and in the case of some contract archaeology situations).

> *"For osteologists, there remains the problem of recognising purposeful intergroup conflict in prehistory. The chance of finding an unambiguous indication of warfare – a massacre of many people - is remote. These incidents occurred infrequently, and the remains must have been deposited in a way favouring bone preservation, typically through*

intentional burial... Reliably differentiating combat fatalities from other traumatic deaths requires the use of multiple kinds of evidence, including the overall proportion of such deaths in a cemetery sample, the number of fatalities in separate incidents, the nature of perimortem damage to skeletons, and a consideration of the archaeological context. The existence of any such evidence, however, demands close attention because problems with recognizing violent deaths ensure that lethal trauma will be underestimated, even in the best of circumstances"

(Milner.1995:237).

Iconography

The kind of iconography that may suggest warfare is remarkably rare in Northern European prehistory; parallels are generally drawn in the literature with examples from Egypt or southern Europe and from prehistoric or primitive examples from Africa, America or Australasia. Interpretation of prehistoric art is a subjective and complex activity, and in order to assess its meanings for an understanding of warfare, it is worth briefly discussing the natures and forms of interpretative strategy that may be applied to it. There has been a trend, in the study of primitive representations, to focus on the practices of shamanism and hallucinatory drugs as explanations of non-iconic symbols - a 'neuropsychological' approach to interpretation (Pearson, 2002).

'Art' has, undeniably, a transforming function, but not all examples of primitive representation need be the result of chemically or psychologically altered states. The practice of using forms of visual 'shorthand' to convey meaning is common in modern society - road signs are a good example. Faced with the need to convey information through the intransigent and labour-intensive means of pecking away with a stone at a lump of granite, it seems likely that the use of shorthand symbols to convey complex meaning might develop rather quickly! Another point, often ignored, is that not all symbols have to *mean* something special. Abstract or geometric patterns, or descriptive scenes, may simply be used as decoration, with no greater significance than is inherent in the desire to demonstrate ownership, influence or use of a particular place by transforming its appearance to one's own taste - to 'wallpaper' a site.

It seems sensible to adopt the attitude that a range of interpretations, including the pragmatic, may be possible for prehistoric art. "Prehistoric artists, like those of more recent times, were people like ourselves and their art can therefore be expected to reflect every facet of life " (Bahn, 1998: 246). Bahn actually goes a little further, considering that "literal interpretations of prehistoric art are far safer - and at worst may be incomplete rather than wrong" (ibid: 221).

There are a series of stages through which any form of artistic interpretation must pass: the identification of the image, the location of the image in time and place, the recognition of the codes of meaning inherent in the image, the understanding of relationships between images, composition and meaning, and the comprehension of the meanings symbolised (Morphy, 1989),

Discussion of San rock art has also shifted from assumption of the realistic to speculation on the shamanistic dimensions of meaning, although "This is not to deny that the San witnessed and were actively involved in the various forms of conflict depicted in the art. Indeed, the paintings were presumably inspired by events the artist either witnessed or took part in." (Campbell, 1986:265), an opinion based on the accuracy of the depictions of weaponry.

That 'primitive' art may be sophisticated, abstracted and multi-layered is argued by Garlake (1987) and in North America, too, this sophistication is recognised. In a discussion of Blackfoot art in Canada, Klassen has noted both iconic and narrative elements (Klassen, 1998:44-5), some of which are used to record the war honours and heroic deeds of individuals, and utilise conventions of a few lines to represent recurring themes such as the number of horses captured in a raid. This use of conventions is also recognised in Pueblo art, where a depiction of a pair of bare foot soles represents an ideogram for 'battle' - the sense being 'stripped down for action'. Similarly, simple triangular shapes represent arrows, with the further sense of 'conflict in which arrows are used'. In dealing with multiple layers of representation, we need to be able to speak the appropriate language.

"A geometric system where there is a relatively arbitrary relation between signifier and signified, form and content, requires an interpreter who is already familiar with the code for the meaning to be grasped, whereas an iconic image, one based on look-alike criteria, may be interpreted at some level by someone who has never seen the image before"

(Morphy, 1989:2).

For the archaeologist, lacking direct access to the languages concerned, learning to interpret the content depends on what have been characterised as *informed* methods (based on ethnographic or concrete interpretation) and *formal* methods (based on physical context and content) of deduction (Taçon & Chippindale, 1998:6-8).

Archaeological contexts for iconographic messages vary in form and the opportunities they offer for informed interpretation. A particularly fine piece of apparently 'narrative' art is that found on the bronze scabbard from grave 994 at Hallstatt, dated to 400-350 BC, which depicts four mounted warriors with spears and three foot soldiers, equipped with spears and shields; one of the horseman is spearing a fallen enemy; in another scene, two men struggle on the ground. "This may record either a real event, perhaps one in which the dead warrior took part, or possibly a mythological battle tale" (Megaw & Megaw, 1989:81). Either way, it clearly demonstrates a familiarity both with combat and with the artistic representation of combat.

The use of iconography to convey political messages has been noted in Egyptian art, often combined with apparently narrative scenes depicting violent acts. Among many other similar examples, the Palette of Narmer shows the king smiting the head of a kneeling captive, demonstrating the triumph, power and legitimacy of the royal status (Filer, 1997). Many scenes shows kings leading armies, again demonstrating their power, strength and courage. To what extent these portray what actually happened is debatable, of course - political allegory is propaganda-based, and kings often have themselves portrayed in heroic form even when they have never been near a battle.

Kristiansen (2001) has argued that similar imagery may be recognised in Scandinavian Bronze Age depictions of the rounded caps and 'ceremonial' axe-bearers that may have accompanied chiefs. These symbols are found in a variety of forms - as hilt pommels, as *tutuli*, and as rock carvings. They are often twinned, which he infers as representing twin chiefs - a warrior chief and a priest-king, a social development resulting from the introduction of new weapons, a change in the style of fighting and the emergence of warrior aristocracies. In this scheme, the rounded cap may be a padded protective piece of headgear which assumes some of the status symbolism of a crown.

Many writers have recognised the usefulness of prehistoric art in the identification of forms of weaponry and armour and in demonstrating the distribution of stylistic typologies of artefacts (e.g. Bradley, 1998; Briard, 1978; Osgood et al, 2000). In particular, indications of padded organic armour have been suggested for Scandinavia, where square bodies and thickened calves may represent such protection (Coles, 1990) and for stelae in Northern Italy and Switzerland *(Figure 6)*, where patterns in the carvings may also suggest padded jerkins; it has also been suggested that the ubiquitous lozenge motif, found, for example, at Bush Barrow and on Beaker pottery, may reflect such equipment (Osgood et al, 2000). Analysis of rock art in a number of locations in Italy, Spain and the Mediterranean islands has led to the identification of what may have been standard panoplies of arms; these depictions are generally considered to be of Bronze Age date, based on the type of weaponry illustrated. Some individuals are shown with a sword, shield, axe and helmet. Most scenes show only two or three combatants. Do these represent weapon display and ritual combat rather than war? (Osgood et al, 2000). Nordbladh (1989:331) certainly believes that the former is the case for Scandinavian rock carvings.

Correlations between weapons found in graves and the rock art of Southern Scandinavia offer some complications to this view. Some 95% of Danish Early Bronze Age male graves were found to contain close fighting weaponry; Kristiansen's analysis of sword forms has led him to suggest that the evidence represents an hierarchical society where fighting is frequent in the upper echelons, for status and power perhaps - but that the swords of the highest level were unused, symbolic items. The very top stratum consisted of leaders (political or religious) of cohorts of warriors who did the actual fighting (Kristiansen 1983, 1984). This may explain why the warriors shown in the rock art are generally formulaic and anonymous (*figure 7*).

Some rock art in Scandinavia shows larger bodies of armed men, although not actual battles. Where armed men are shown in conjunction with ships, there are usually only a few figures shown. Does this really indicate that fighting was in the forms of duels and that wider or larger scale raiding did not happen? The problem facing interpretation is again one of understanding artistic convention and meaning; it is perfectly reasonable to suggest that the material of the representation (hard stone) and the effort required to utilise it would have led to the use of visual shorthand, so that a boat plus two or three armed figures *stands for* a generally understood meaning of a raid or sea battle. In the same way, the absence of depictions of victor and vanquished can be explained by the understanding that the pictures were not intended as memorials to a particular encounter, but rather as illustrations of that encounter or such encounters in general.

Sardinian bronze figurines include some 80 warriors dated between the 11th and 7th centuries BC - most seem to show protective clothing, possibly of leather; some have long skirts with bosses (maybe bronze discs), pectorals, spiral neck-guards, knee-guards or greaves. All but one have helmets, half of them horned; many have round shields with a central boss. The main weapons depicted are the war-stick, bows and arrows, daggers and swords. It seems that there were different sets of equipment - about a quarter have helmet, doublet, round shield and war stick, and often greaves and dagger. Nearly half have a helmet (generally horned), doublet and bow, and most have a pectoral, sword or dagger and greaves.

Such 'sets' of equipment may represent different status of warrior, or different types of troops. Similar 'sets' are identified in Iberian rock art examples of the Late Bronze Age by Bradley (1998). The implications for pictorial representation as a record in changes of style of fighting are also recognised by Osgood (2000). Scandinavian carvings tentatively dated to around 1000 BC appear to show chapes on scabbards, suggesting an increased use of horses in fighting. The Late Bronze Age Strettweg chariot figures include mounted warriors with shields, spears and crested helmets. It has been suggested that two of most important aspects of Bronze Age iconography are the appearance of horses and boats as clues to types of fighting (Osgood et

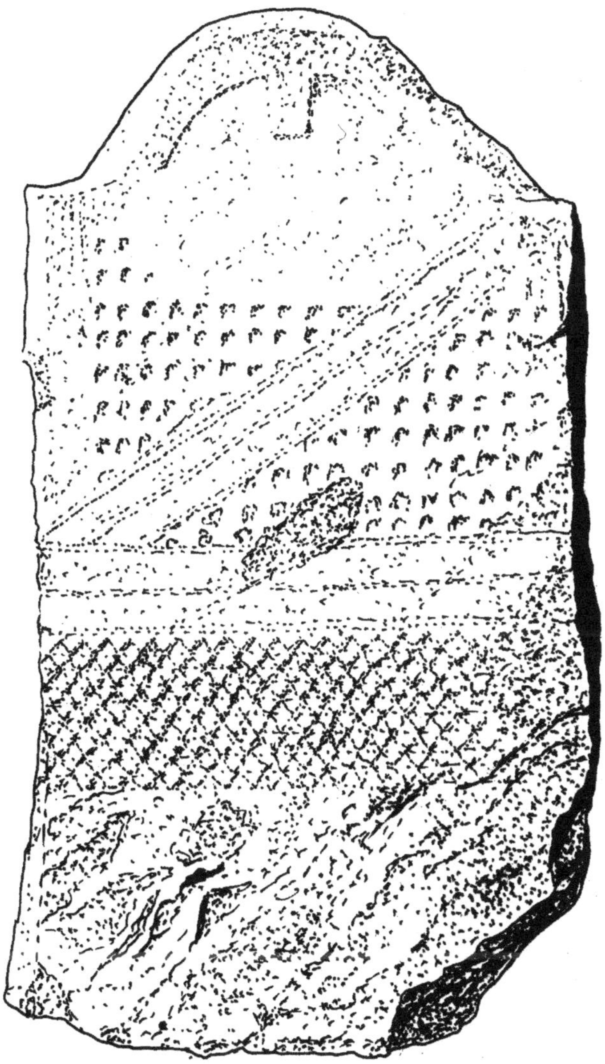

Figure 6: Statue-menhir from Petit Chausseur, Sion, Switzerland [after Carman & Harding (eds) 1999:160]

al, 2000: 145). The Roos Carr figures from Humberside (which may date from the Late Bronze Age) seem to represent naked warriors armed with clubs and carrying round shields, possibly standing in a raiding boat (*Figure 8)* (Osgood, 1998).

The *context* of the representations remains problematic. Rock art has been associated with ritual activities, even forms of pilgrimage in the landscape. Some examples may be memorials - particularly, perhaps, the statue-menhirs of Iberia and the Alpine regions. Representations of weaponry may indicate the vicinity of weapons-hoards, or serve as symbolic weapons-hoards (Bradley, 1998:135) or as a 'display of trophies' (ibid: 141).

Elsewhere, what might appear to be examples of armour or weapons are interpreted in completely different ways - the 'corselet' shapes found on stones at, for example, Neolithic sites at Les Pierres Plates and at Locmariaquer, Brittany, are referred to in one publication as "somewhat complicated figurations derived from the mother goddess perhaps with multiple breasts" (Briard, 1997:47). The designs of Gavrinis of a similar period are similarly described as evocative of the mother goddess, despite the very clear representations of axes, both hafted and unhafted. But "La hache est aussi symbole du feu, de la foudre; c'est une symbole céleste; il peut être associé au serpent, symbole de la terre....Ce symbole s'affirme donc comme un incontestable emblème du pouvoir" (L'Helgouac'h, 1998:10) so perhaps any interpretation one chooses is equally valid.

A form of symbolic representation rarely cited in discussions of evidence for war is the war memorial. Whilst not in every case offering 'iconography' as such (although many do), these structures or monuments

Figure 7: Rock carving from Tegneby, Bohüslan, which seems to show an armed confrontation between mounted warriors (after Randsborg, 1995:a)

may form a visual symbol in the landscape that requires interpretation. Numerous examples of statues and plaques exist that commemorate the fallen of the First and Second World Wars - in nearly every town and village of Britain, France, Belgium, the Netherlands and many other countries, bronze and stone installations have long outlived the actual conflicts, and will probably survive for centuries to come. The Romans were equally keen on large and imposing monuments to commemorate particular campaigns and victories. Triumphal arches have proved to be of enduring popularity for this purpose - Paris has the Arc de Triomphe, London has Marble and Admiralty Arches, and there are many more across the world.

Evidence for war memorials in the prehistoric period would necessarily lack defining inscriptions; nevertheless, there are some examples of features which may well be interpreted as this kind of monument. The Hjortspring and related Scandinavian deposits certainly seem to be examples; it has been suggested that some Bronze Age tumuli, particularly those which do not seem originally to have included interments, may in fact be cenotaph monuments, commemorating fallen warriors whose bodies could not be recovered for burial at home. This has certainly been suggested as an explanation of the central barrow and six others of the West Heath (E.Sussex) group, one of the Iping group and the bell barrow at Deerleap Wood, Surrey, (Drewett et al, 1988) and this (combined with the nineteenth century example of the Butte de Lion at Waterloo) raises speculation about the original purpose of a monument like Silbury Hill.

Among the most convincing finds of the 'war memorial' type are the macabre sites of Ribemont-sur-Ancre and Gournay-sur-Aronde in Belgic northern Gaul. These sites comprise rectangular 'sanctuaries', enclosed by timber palisades within larger enclosures. At Gournay-sur-Aronde, excavations have recovered some two thousand pieces of armour and weapons, many of which seemed to have been arrayed on gates, poles or a platform, and representing perhaps a century of warfare. At Ribemont-sur-Ancre, ossuary deposits representing up to 1,000 headless warriors, many of the bones burned or crushed to extract the marrow, were found arranged in flat-topped 'banks' of bones, bone-filled ditches (*Figure 9)*, and what appears to be a wooden platform upon which perhaps dozens of decapitated skeletons were posed with their weaponry (Brunaux, 1999). The excavator has interpreted the remains as being those of defeated enemies, deliberately mutilated and collected. The absence of skulls is thought to confirm the practice of carrying the heads of defeated enemies into battle. "The sanctuaries seem to have served a double function, martial and cultic. The Gauls designed them to celebrate victory in war and intimidate potential enemies while pleasing the gods of the underworld, whom they believed made them great warriors.....The human remains and weapons at Ribemont are examples of the dedicatory heaps of booty piled up in consecrated areas that Caesar mentions in his account of the Gallic War" (ibid:57).

Other hoards and monuments may fall into the category of memorial, as may iconographic representations of conflict,

Figure 8: The Roos Carr figures (after Osgood, 1998:35)

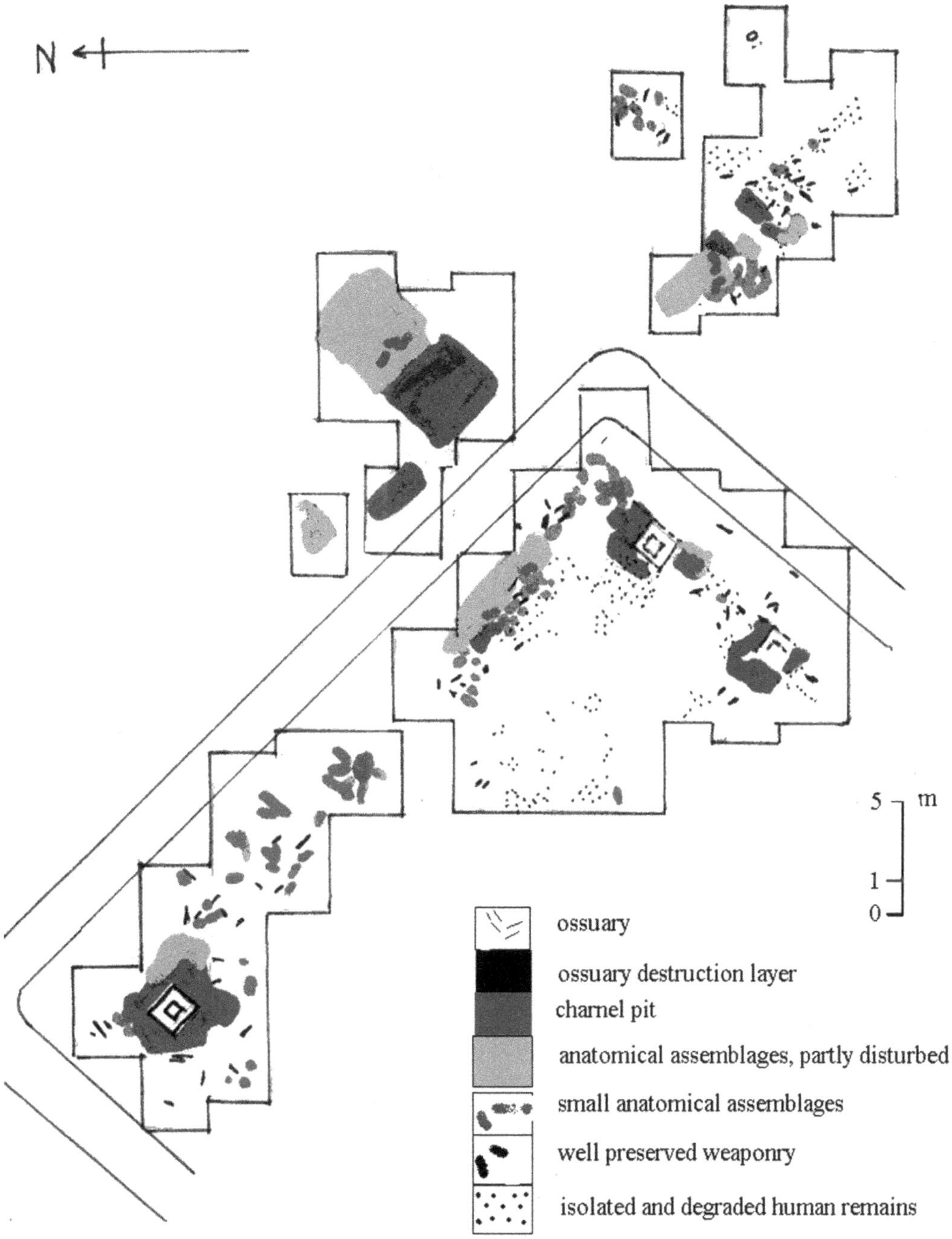

Figure 9: Ribemont-sur-Ancre, Somme (after Brunaux et al, 1998)

warriors or weaponry. In this context, we need not rely on exact accuracy of depictions of tactics or *materiel*; memorials often show images of victory or suffering unrelated to the actual period or circumstances of the conflict being commemorated. A brief survey of local war memorials commemorating the 1914-18 and 1939-45 Wars has produced a list of incorporated symbols that includes Graeco-Roman armour and laurel wreaths, Napoleonic period lances and cannons, Greek-style female winged Victory figures, medieval tilting helms, and naked warriors engaged in one-to-one duels with swords and shields, as well as reliefs and statues that show modern combat clothing and rifles, aircraft and tanks. As with symbols related to royal power, war memorials have a tendency to hark back to eras with more glamorous connotations, and there is no reason why this should not have been true in the past.

It is abundantly clear that whatever the interpretations, artistic representations can on their own provide very little positive evidence for the study of prehistoric warfare, although they seem to indicate a familiarity with many of its symbols and prerequisites. It is nevertheless possible to suggest that some forms of representation or symbolic structures *may* have been intended to record actual arms, actual conflicts or actual political power through force, either as realistically as possible within the artistic conventions of the period, or as emblematic (even archaic) encoded representations, or as memorials, deposits or sanctuaries offering a form of symbolism that would have been readable in their contemporary landscapes.

Conclusion

The above discussions seem to indicate that the most commonly cited forms of evidence for the recognition of warfare in prehistory are all problematic, at best often ambiguous, and frequently too insubstantial to be used to make firm interpretations. Many organic or lithic weapon forms are simply not recoverable, and others are rarely preserved in contexts that can inform research about their use in fighting; fortifications are complex in their meanings and the intentions of their builders, and often have been much altered from their original form. Skeletal evidence is extremely rare and evidence of weapons trauma has many different potential causes that may be unrelated to warfare. Interpretation of iconographic evidence is often very subjective, and it is not possible, in most cases, to understand the symbolic and cognitive meanings that may have been familiar to its contemporary creators and audience.

It is therefore necessary to consider what other resources may be available to archaeologists in attempts to study the phenomenon of prehistoric warfare. A starting place may be an attempt to clarify the forms of conflict that may have taken place.

Chapter Three - Tactics and Strategies

The way a war is conducted - its organisation, tools and procedures - may, in retrospect, offer an interpretation of its aims. Raids for booty may be conducted differently from revenge attacks, and assaults aimed at encouraging the offering of tribute are likely to take a different form from those aimed at territorial expansion. There does exist a certain amount of evidence to illuminate this aspect of hostilities, partly drawn from artefactual material, and partly from forensic investigation of geographies of battle or attack. Even more useful is the evidence of the immediate effects of violence - dead bodies, burned villages and the like, demonstrating the scale, intensity and destructive imperative of the attackers.

The tactics of prehistoric conflict may be more accessible for study than the strategy, assuming that there was any degree of strategic planning. Ways to recognise tactics may include the evidence of actual attacks, comprising damage to structures, distribution of weaponry and episodes of destruction, and the treatment of enemies revealed by study of skeletal remains. We may also look at the design of boundaries, settlements and fortifications, and consider ethnographic evidence.

Tactics will depend on the aims of conflict, the size, composition and structure of the troops, and the degree of complexity of the societies involved. Warfare is an infinitely variable activity; for the purposes of attempting to identify forms of prehistoric conflict it seems advisable to create several fairly simplistic scenarios, and to investigate the possibilities for evidence and interpretation they offer. At the same time, it should be remembered that there *are* an infinite number of scenarios, aims, organisations, and other variables that can be imagined for the past; it would be impossible to go through all the alternatives in the present discussion.

The scenarios that can be identified in some of the archaeological material available include:

- sporadic spontaneous battles or assaults, often associated with 'revenge' or 'sport' warfare;
- low level long term endemic warfare associated with raiding for portable booty;
- individual larger scale assaults on fortified positions associated with raiding and/or status competition;
- assaults associated with possible territorial expansion or cultural or ethnic conflict; and for comparison -
- large scale territorial expansion and conquest (the Roman annexation of Northern Gaul and Britain).

It is recognised that the lower levels of hostility can easily escalate into higher ones, or, indeed, decrease in intensity over time. Within these scenarios, it will also be possible to discern the six tactical systems proposed by Otterbein (1989). Of these, four are systems that operate in parallel to the enemy forces: troops arrayed in lines, utilising cover and projectile weapons; troops arrayed in lines using projectile weapons without cover; troops in lines at close quarters, using hand to hand weapons; and troops divided into specialised units under co-ordinated direction. Two further systems are not parallel: ambushes in which the enemy is surrounded; and ambushes where the enemy is led into a trap. Any of these systems could be utilised in the scenarios suggested.

Sporadic or spontaneous tactics

Ethnographic accounts may offer some illumination of the tactics employed by societies with pre-modern technologies and organisations. Sporadic, spontaneous battles are the form associated with modern non-western societies such as those of New Guinea and, of course, Chagnon's Yamamomö (Chagnon, 1968). The tactics associated with this very basic mode of conflict anthropologically include short duration of active hostilities, a low level of command and discipline, a great reliance on surprise attacks (Vayda, 1967) and sometimes, the mutual agreement to fight (von Clausewitz, 1976 [1832], 245; Keeley, 1996:60; Keegan,

1993:296). Claessen records that among the Hurons of nineteenth century North America, any person desiring to instigate hostilities could assemble a force [a situation that he likens to Julius Caesar's description of North Gaulish prominent warriors who had the power to announce a raid and invite anyone who wished to join in to do so (Claessen, 2000:105)].

Sneak attacks and ambushes are often the preferred procedure in these types of assault, under cover of darkness, at dawn, or aimed at isolated dwellings, individuals or small groups of people, often non-combatants. Generally, only a few people may be harmed in any one incident although the cumulative effect may be serious.

> *"Raids characteristically kill only a few people at a time; they kill a higher proportion of women than do battles or even the routs that follow them; they kill individuals or small groups caught in isolated circumstances away from major population concentrations; and because the victims are outnumbered, surprised, and often unarmed, their wounds are often inflicted as they try to flee. Archaeologically, this pattern will thus be evidenced by four corresponding characteristics: burials of individual or small groups of homicide victims; women as a high proportion of victims; burials sometimes located away from the major habitation zones (although raid victims were recovered and buried in the usual cemeteries); and evidence that most wounds, even on adult males, were inflicted from behind"*
>
> (Keeley, 1996:66).

Individual assaults on stragglers or strangers, or sporadic raids for booty can lead to cycles of revenge escalating into skirmishes and even massacres (Chagnon, 1968). In some cases, an agreed and expected aim of the attack is to destroy not only the guilty party but also his family and neighbours, as recorded by Balikci (1970:184) among the Netsilik Inuit, and by Vayda (1960:92) among the Maori.

Possible archaeological examples of this form of attack include, it has been suggested, the assault that resulted in the deposition of skulls at Ofnet, Bavaria, in the Mesolithic c.6500BC.

Pathological examination has shown that most individuals suffered blows to the occipital area, and some had suffered earlier, similar, assaults. Frayer (1997:209) suggests that "it seems very likely that the individuals in each pit represent a mass burial resulting from a single incident" which may have been the type of revenge raid described above, especially in view of the high proportion of children and females and the fact that they had been decapitated, perhaps in the acquisition of trophies (Thorpe, 2000). Other such examples occur at Korsor Nør harbour settlement in Zealand (Thorpe, 2000), and the Linear Pottery settlement at Talheim where a mass grave of at least 67 individuals, half young adults and half adolescents and children, was discovered. Dating from about 5500-5000 BC, the people had suffered blows to the skull from shoe-last adzes or had been hit by arrows, and had been thrown without ceremony into a ditch near the gate of a double oval enclosure in one incident; further skeletons had been thrown into a well. Many of the skeletons had suffered damage from animal scavenging (Vencl, 1999:61-2; Bahn, 1996: 48-49). At Herxheim (Rheinland-Pfalz) some 334 bodies have been found thrown into ditches without ceremony, and groups of the bodies of victims of violence that include women and children have also been found at the Jungfernhöhle Cave in Upper Franconia, at Barleben in Germany and at Ruzindol-Borov, Slovakia, all dating to the Neolithic (Vencl, 1999: 62-64), and the multiple burial of 12 individuals at Wassenaar in the Netherlands dating from the EBA/MBA transition around 1700BC, mentioned above.

These examples display many of the characteristics associated with discussions of conflict - atypical burials, scavenger damage, and choice of victims - further illustrated by examples in the following section.

Endemic warfare and raiding tactics

Sporadic raiding can easily develop into feuding and thus long term endemic warfare. This form seems to have occurred in prehistoric North America in a number of locations and periods. Chatters (1989:243) distinguishes between feuding and warfare:

> *"War is defined... as organised violence conducted in the name of the community as a whole, in which combatant groups act as units in organised fighting. Feud is prolonged conflict between members of intimate groups or of politically distinct communities that is not authorised by the community as a whole. It is motivated by a desire for revenge or reprisal and each act triggers a subsequent act in a cycle that may last for generations. In a feud, the level of damage inflicted by each act of violence is that which is considered sufficient to avenge the previous act.. "*

Thus war involves groups of young adolescent males facing each other belligerently; it follows that most wounds will be sustained to ventral body surfaces and victims will be buried relatively formally close to home or to the site of battle. Feuding parties would be less formally constructed and might be more likely to employ covert tactics against any victim that presented themselves. These victims may therefore be of any age or gender, and suffer wounds to dorsal surfaces or present evidence of 'defence wounds', and their bodies might be found in remote locations. This definition is perhaps rather too simplistic. When feuds, raids for plunder and acts of revenge escalate into longer-term and more lethal conflicts, the tactics associated by Chatters with feuding are likely to be retained; formal face-to-face battles are more likely to be restricted at this level to forms of 'sport' fighting. They are not representative of true hatred or rivalry between individual combatants, a phenomenon of formalised battles that has been seen in many more modern battlefields, such as the trenches of World War One, where hostilities could sometimes be put aside between individual British and German soldiers in particular circumstances.

Chatters gives a number of examples of the probable victims of attacks, possibly resulting from endemic warfare, from 14th century AD Oklahoma (1989:244-245). It is recorded that half of victims in these and other graves were likely to have been non-combatants; wounds were primarily dorsal; and clustering of dorsal wounds in vital areas suggests assailants were close to their victims (ibid:246). Graves were often isolated and many bodies showed scavenger damage, suggesting time had elapsed between the assault and the recovery of the body by their communities.

> *"The conclusions that can be drawn from this evidence are:*
>
> 1. *Conflict did not consist of organised fighting between warrior groups. Rather it consisted of attacks on any individual of the enemy group. There was no preference for killing combatant males or even adults.*
> 2. *Sneak attack and ambush were preferred over direct confrontation, particularly when able-bodied males were among the intended victims.*
> 3. *The assailants further avoided personal risk by attacking individuals or small work parties who were separated from their fellows"*
>
> (ibid:246).

A well-known example of supposed endemic and ultimately attritional raiding warfare is the Norris Farms #36 cemetery in Illinois mentioned in a previous chapter. Many of the skeletons showed evidence of pre-existing debilitating conditions including nutritional deficiencies, infections, iron-deficiency anaemia in up to 85% of individuals, and many cases of what has been identified as pulmonary tuberculosis.

> *"The vulnerability of adults whose work would have exposed them in positions some distance away from the protection provided by the settlement is reflected in the age-at-death distribution of victims. It seems that even people with disabilities were called upon to range widely in subsistence activities. Individuals with some form of prior dysfunction are represented among the carnivore-damaged and disarticulated skeletons. These remains are assumed to have belonged to people who died in remote locations .... Feelings of insecurity resulting in excessive caution when conducting subsistence practices would influence the success of food-procuring activities. This disruption might find expression in the size of the area regularly traversed by foraging parties, the number of people required to conduct resource-acquisition tasks reasonably safely, and the need for sentinels, which drains an already depleted labour pool. Regular access to resources located far from the protection provided by the settlement would be especially problematic... Isolated and, hence, vulnerable work parties would be particularly tempting targets especially whenever their movements are highly predictable because of the restricted spatial distribution of resource harvested only at certain time of the year."*
>
> (Milner et al, 1991, 589-591).

Endemic raiding was a feature of Greek prehistory as much as elsewhere, although carried on with more developed troop and organisational management. Annual raiding was, at times, almost predictable, either while the crops were in the field or more usually, just after harvest, when all the produce was gathered together in the villages and farmsteads, and so easily plundered. In some cases, deliberate care was taken not to damage the farms too greatly, so that they would continue and therefore be available for plunder in succeeding years (Hanson, 1983).

A few true raiding cultures have been identified by anthropologists, one of the best known examples being the Apache. These people hunted but rarely farmed; instead, they regularly raided their neighbours (the Navajo, Hopi, Zuni and other farming tribes and Mexican farmers and ranchers) so predictably that in self-defence it became the custom to corral a few animals at some distance from settlements, where they could be taken by the Apaches and violence could be avoided by all parties. Such defensive strategies would, of course, be archaeologically invisible.

It has been argued that forms of endemic and raiding warfare were the norm for prehistoric conflict.

> *"A survey of ethnographic and anthropological sources showed that, first of all, violent conflict is the rule rather than the exception in all societies. Secondly, in many tribal societies warfare is endemic, either as raids or in a more ritualised form, and, thirdly, these tribal types of warfare do generally not leave conspicuous archaeological traces: they do not involve defensive structures, specialised weapons and armour, large-scale destruction, specialised warrior groups, etc. In view of these observations it seems permissible to use the little Neolithic and Middle Bronze Age evidence to assume that armed conflicts or tribal warfare were*

endemic throughout the Neolithic and the Bronze Age. This form of warfare was the logical basis for the more visible forms of warfare of the more complex societies of later prehistory."

(Louwe Kooijmanns, 1993:18).

Tactics associated with prepared fortifications

Victims of endemic violence may respond by building fortifications, and the tactics of assaulting these are rather more visible. The risks associated with attacking prepared defences suggest that the aims of the attackers are rather different from the casual raiding for booty, women or revenge posited for the preceding scenarios and evidence. Whilst raiding for booty may still be a part, even a major part, of the motivation, status competition between rival groups seems to be indicated as an important additional impetus. This, and the tactics required to carry out the attack successfully, begin to imply leadership and organisation at a slightly higher level. This may be a development out of previous more informal raiding culture and practice, and it might also lead to later features of warfare such as the training and employment of specialist troops, command structures and prepared logistics.

Ethnographic descriptions of assaults on prepared defences include those carried out by Maori warbands. Maoris constructed defensive villages, the '*pa*', surrounded by one or more lines of ditches and palisades, some equipped with fighting platforms, and narrow gates. Within the enclosure at the highest point was usually the chief's dwelling, which formed a kind of citadel, and was used as a refuge and place of last stand during attacks. Because of the strength of these villages, tactics for war tended to revolve around sneak attacks, stratagems such as approaching disguised as wood collectors, use of scouts and attacks under cover of darkness (Vayda, 1967). Raids often made use of canoes, and the whole village community combined to support the warbands, which typically consisted of about 140 men. Battles were never of long duration, not more than two days, but hostilities could occur over a much longer period - even generations. The war chief was usually the eldest male, although he could nominate another, but the organisation of the fighting was lacking in complexity - leaders rather than generals. Trophy gathering and display was part of the nature of these fights, with trophy heads being mounted on palisades. Such remains were denied deposition in tribal sacred places, and offered a visible form of intimidation to potential enemies. Prisoners were usually summarily killed, but some were sometimes retained as slaves or to be slaughtered, and indeed eaten if food was scarce, later. The main aim of the assault was the acquisition of portable booty and the demoralisation and massacre of the enemy, including non-combatants. Warbands could be in the order of up to 500 men if several villages with contiguous territories decided to form alliances, and although territorial expansion was rarely a primary aim, a series of successful attacks could have the effect of driving communities away, in which case their lands were adopted by the victors. The displaced community might then either colonise virgin territory if available, or commence hostilities against more distant communities to gain new lands (Vayda, 1967: 375).

Attacks surmised at a number of Neolithic enclosures in Britain and Northern Europe may have been similar in aims and execution. An increasing number of Neolithic sites show evidence for defensive building (although there is some argument about whether these features were intended as such or were a symptom of an increase in social complexity and status marking), and some seem to have evidence of violent assault. Like the Maori 'pas' , these seem to have utilised raised positions and in some cases involved the construction of ditches and palisades, with restricted entrances, and possible fighting platforms. Mercer has summarised the evidence from Crickley Hill and concludes that "This enclosure seems to have been the target of intensive and probably tactically marshalled archery" (1999:152). That archery was an organised tactic of assault is supported, he believes, by the design of the defences, an opinion shared by the excavator:

> *"The enclosure had quite obviously been defended against archery attack and it is highly likely that it was built with this intention, for the low palisade formed no more than a breastwork. The ditches were presumably designed to break up and slow down an assault, and the low bank, or rather platform, would then serve as a killing-ground, at point blank range, against aggressors clambering out of the ditch. It is worth noting that despite ditches slighter than those of phase Id, the bivallate interrupted ditch enclosure would, on this interpretation, be a defence in depth more effective than the final rampart even though their (the latter) larger size gives them a greater appearance of strength"*

(Dixon 1989:82).

Mercer also excavated at Hambledon Hill, Dorset, where evidence of an assault involving fire and the use of arrows was found, and at Carn Brea in Cornwall he believes there was "an assault involving massed archers followed by frenzied destruction" Mercer, 1999:153), possibly perpetrated by the inhabitants of the nearby Helman Tor who may have been status rivals. Further evidence of destruction by arrows and fire has been suggested in Northern Ireland at the Donegore Hill enclosure where there were several burned houses, possibly destroyed in an attack by occupants of nearby Lyles Hill, and at the Billown enclosure, Isle of Man, where concentrations of arrowheads are held to be evidence of an attack (Thorpe, 2000).

Ferrill claims that Early Neolithic Spanish Levant paintings represent the use of column or line formations, and differentiation of leaders. "The appearance of the column and the line, which imply command and organisation, is synonymous with the invention of tactics." (1985:21). The dominant tactic, in Britain as elsewhere, seems to have been an organised, and possibly directed, long-distance arrow storm followed, once the defences had been 'softened up', by close quarter fighting and destruction by fire. Investigations at Chalcolithic defended sites in Spain show similar tactics. Analysis of the architecture of these sites ".. can provide inferences on the frequency and nature of attack, the types of weapons and tactics used, and the outcome of particular events" (Monks, 1997:9), and from " the structural evidence, it would appear that projectiles were the principal weapons and had reached a high level of efficiency" (ibid: 22).

There is evidence of burning and subsequent abandonment at these Spanish Chalcolithic sites - not necessarily due to war but possibly significant. Massacres had occurred -

> *"There are several different types of warfare represented .from an intense,indiscriminate possible massacre at one site [San Juan ante Portam Latinam, Laguardia (Alava) dolmen....], to more sustained, regular patterns of violence at others. Almost all the injuries are the result of attacks using projectiles, usually thrown at a distance....Some individuals showed the presence ofearlier injuries and wounds, often with some degree of healing or treatment, proving that individuals had survived previous attacks." (ibid:24).*

Confirmatory evidence for hostility in the region may come from cave paintings portraying conflicting groups and archers and weapons such as the battle scene at Les Dogues in Castellon, Eastern Spain which shows two highly distinct bands of warriors opposing each other (Beltran, 1982:44).

"The presence of dense patterns of settlement, sometimes clustering around a fortified site, implies that conflict may not have been between neighbouring groups, but perhaps between different regional areas or valleys" in the Spanish Chalcolithic (Monks, 1997:25). There seems to be some evidence for a change from long-range to middle-range tactics in the Chalcolithic and Bronze Ages which may be related to the apparent disappearance of defended sites during the period.

Randsborg charts the appearance of the chariot and helmets in Scandinavia, together with the dominance of swords in aristocratic graves and the absence of weapons in ordinary graves, and the fact that swords outrank spears in hoard deposits in Bronze Age Denmark, and suggests that this represents a tradition of aristocratic duels rather than larger scale conflicts, up until about 1300 BC.

> *"While the metal-tipped spear and/or lance were perhaps the most important innovations, in use throughout among both high and low, the costly sword came to signify the highly personal, almost mystical and often very well-made weapon with an aristocratic air about it. The sword was, for many centuries to come, to symbolise personal standing, while the lance was rather a symbol of adulthood and authority (cf. earlier the axe)" (Randsborg, 1995:146).*

In the same period, defended sites seem to be absent in the British Isles too; it is possible that there was a widespread tradition of conflict resolution by duel, possibly at least partially ritualised, in Northern Europe through most of the Bronze Age. However, raiding must surely have occurred from time to time, and various depictions of boats with armed crews of up to 30 men in Southern Sweden *Randsborg, 1995) may represent an otherwise invisible form of attack, although probably not warfare *per se*.

Different areas seem to have adopted or retained alternative styles of fighting. Depictions on stelae in the Transalpine region show that in the Rhone Valley, Switzerland, warriors were using bows for long-distance offence or daggers for close quarters; on the other side of Alps warrior images had daggers and battle-axes. (Harding, 1999:160-161).

Developments of tactics from the Later Bronze Age

From about 1300 BC onwards, however, another change seems to have occurred. The reappearance of defended or defensible settlements is accompanied by a massive increase in the numbers and types of weapon available, and by inference, a change in fighting tactics, perhaps building on earlier experience.

Randsborg (1995) traces the rise of infantry to the period 1300 to early 1st mill BC, suggesting a remarkable similarity of elements between the Near East, the Aegean, Central and Northern Europe. Common types of flanged sword and dagger blades, metal and other helmets, shield types, bronze cuirasses, and bronze greaves appear, suggesting the employment of a new form of mobile light infantryman. He further suggests two models of early Germanic warfare that may stem from this period:

- Model I - Raids: Attack by limited elite forces, especially boat crews, countered by large or even very large (regional) militia armies, perhaps supported by their own elite forces and/or by a small cavalry force. The larger force will usually win because of the defensive superiority in men and supplies.
- Model II - Battles: Elite forces with a militia army and train in the form of a tribal unit and its resources of people, animals, food, tools etc. This would be met by the home militia army with its possible elite forces. In this model the outcome hangs in the balance.

The beginnings of formal armies, then, might be traceable to the Late Bronze Age in Northern European contexts.

Into the Iron Age, the question of attacks on defended sites and the viability of hillforts as defences has been much discussed (see above). Whilst conceding that hillforts (whether or not on hills) may have multiple functions, it is argued that there is sufficient evidence to show that design elements were incorporated into them which demonstrate an understanding of fighting tactics - funneling and complexity of entrances, design of ramparts and changes in design of ditches etc.

Early hillforts of box-rampart construction with simple entrance structures have been described as 'general-purpose defences' and the gates would have worked as means of constricting approaches and probably providing platforms for defence.

> *"The weapons envisaged by these tactics were clearly short-range weapons: sword, axe, spear and presumably such occasional stones as lay to hand... tactically they are the heirs of the Late Bronze Age tradition of weapons and of 'close-combat' fighting, the longest range weapons being the spear and perhaps the throwing axe"*
>
> (Avery, 1986:222-223).

Avery believes that this represents the end of the Bronze Age world of leadership by warrior elites and simple chiefdoms.

Later developments associated with the various forms of dump ramparts with funnel or inturned entrances suggest different tactics. Julius Caesar records that at hillforts in Gaul in the first century BC, the normal method of attack was to drive defenders away from the rampart with the use of slingshots, then using shields for cover, charge forward and burn down the gate (*de bello Gallico*, II.6.2). Avery suggests the development of complex gates was a response to such tactics (Avery, 1986:223). In turn, attackers would need to develop new tactics themselves, adding levels of sophistication in tactical thinking, weaponry and training.

> *"They make demands upon the commissariat of the attacker, who must bring up skilled slingers to mount a heavy barrage (and stand-ins for any injured by the defence) and supplies of sling-stones. 'Stoning and fire' would hardly succeed first time at every site: supplies must be sufficient for repeated attempts. Sling-stone supplies would form relatively static targets... They would be vulnerable to a sally by defenders... If attack by stoning and fire is to succeed, the attackers must be in effect able to control the approaches to the fort and mount almost a primitive form of siege" (ibid: 225-226).*

Warfare of the type described by Avery is no longer 'heroic' - it is on its way to becoming the domain of professional troops. We do not need to think of particularly massive mobilisation, however. It is worth remembering that according to the Laws of King Ine of Wessex in the late 7th century AD, a group of seven men up to no good were merely thieves, between seven and 35 were a band and a party of over 35 men was an army (Thorpe, 2000).

The analysis of the deposit within the bog at Hjortspring in Denmark "allows for the reconstruction of a mobile elite fighting unit of four commanders (mail, narrow shield, sword, bayonet-lance) and eighteen ..common warriors (broad shield, one spear, one javelin - half with a broad, half with a narrow head). Half of these warriors would also have had a javelin. The composition of the force reveals both uniformity and a specialisation of tasks" (Randsborg, 1999:195). Whilst admitting that the find may be atypical, he claims that it clearly shows a 'small, barbarian edition' of the phalanx formation of the contemporary Mediterranean world. The weaponry found, according to Randsborg, looks like an army, consisting of a minimum of three or four ships, each having a steersman and possibly an assistant, two other commanders, all with swords and perhaps a heavy lance each, and 18 warriors, each armed with a heavy spear for close fighting and a light spear for throwing.

The origins of these troop formations are traced to the Bronze Age by Randsborg - analysis of material from Smørumoure, Zealand, another bog deposit, suggests groups of five warriors each with a spear or lance and an axe, with one junior commander with two spears, a lance and an axe, and two senior commanders each with a light and a heavy spear, an axe, a sword and a dagger. This could represent about 50 men in two boats (similar in size to the Ferriby example), capable of forming a front about 100m wide. The tactics employed by such bands could be those described by Tyrtaois in about 650BC "...opposing breast to breast that's how to fight, with the long lance or sword-grip in your hand. You light-armed men, wherever you can aim from the shield-cover, pelt them with great rocks and hurl at them your smooth-shaved javelins, helping the armoured troops with close support" (quoted in Randsborg, 1995).

At Tegneby Mellangården, N Bohuslan, a rock carving shows a clash between two small groups of mounted warriors (four against five) with spears and square shields, with two unmounted men standing behind the riders (Randsborg, 1995) (*see Figure 7)*. It is tempting to interpret this scene as representative of the kind of troop deployment and direction suggested.

Similar evidence of troop composition is lacking from North America, but the attack on the community at Crow Creek around AD1325 demonstrates that the building of fortifications was no guarantee of safety, as it appears that it was while these were being refurbished and rebuilt that the final devastating attack occurred, with over 500 casualties (some 60% of the population) and the burning of all the houses. This village was never reoccupied (Keeley, 1996:68). It is hard to believe that the attackers had not employed tactics such as the use of scouts to be able to choose the most opportune moment for their assault.

Tactics associated with territorial, political or ethnic conflicts

Assaults deliberately aimed at territorial expansion or the destruction of other groups for cultural, political or ethnic reasons, tend to involve greater numbers of combatants, and therefore may be more likely to employ more complex forms of leadership and direction, and more sophisticated and extensive ranges of strategic and tactical actions. The acquisition of portable booty may be only a secondary aim, and there is likely to be a greater level of general destruction associated with conflicts. In Chihuahua, Mexico, the destruction of the settlement at Casas Grandes in AD 1340 may have been a result of this kind of larger warfare; the excavators found "collapsed and blackened walls, suggesting the burning of the city; the presence of breeding turkeys and macaws left to die in their pens; unburied bodies found throughout the city; and the discovery of broken altar stones and figures" (Ravesloot & Spoerl, 1989:131).

The ethnic war in Croatia in recent times produced similar evidence for destruction and abandonment. "War destruction has been significant and has taken a major toll in the agricultural population both in humans and farm animals. In addition, many villages, farm land, buildings and much equipment were destroyed; in other areas agricultural activity has been prevented" (Richardson, 1995:76). The Croatian Army Veterinary Corps reported animals left to die of thirst or starvation in barns and pens. In the Lika region, 50% of livestock had been killed or stolen, and in the Novska region, 734 cattle had died of catarrhal-purulent and haemorrhagic pneumonia preceded by distinct acidosis of the stomach, as a result of abandonment and exposure (Gasparovic et al, 1995).

Territorial strategies may include the forced emptying of regions, either for ethnic reasons or in order to make statements about the military might of a society. Caesar says of the Germanic Suebi "They regard it as the proudest glory of a nation to keep the largest possible area round its frontiers uninhabited, because it shows that many other peoples are inferior to it in military might" (Caesar, tr. Handford, 1951:109). The land is apparently unused and unexploited - the requirement for propaganda and tribal status is pre-eminent. The Suebi had harassed the Usipetes and Tenctheri, neighbouring tribes, to the extent that they were unable to farm their land and were forced to migrate from the region; Suebi society, Caesar tells us, was formally organised for the provision and support of troops forming large armies specifically for foreign campaigns, and military training was normal in their society. According to Caesar, they were capable of putting an army of 100,000 armed men into the field annually.

Once discussion of such large fighting forces is introduced, logistical problems might become more relevant. Randsborg suggests that what Caesar saw as migration of Gaulish tribes (such as that of the Helvetii) might in some cases have been the result of a response to the threat offered by his approach: " barbarian Iron Age migrations may have come about for military reasons. The necessity to launch large forces to fight Mediterranean armies would have taken a large part of a male population away from home. The logistics would have called for additional support from the women, and a migration would have been the result..." (1999:197). However, he does not see this as a feature of warfare in the north. "Large armies of conquest, capable of expelling populations, could not be formed and supported for any longer period, unless whole tribal groups moved along too. Of such large scale population movements we have no evidence from Northern Europe" (Randsborg, 1995: 38).

Responses to large formal armies intent on territorial expansion (with or without ethnic or complex political motivations) could require a far greater degree of complexity in the make-up of both offensive and defensive troop components. Tacitus' descriptions of Germanic fighting tactics suggest something of the specialisations postulated by Randsborg in the variety of weapons used by foot and mounted troops and in the use of such troops in combination and in wedge shaped formations in the battle line. Tactical operations might involve the use of camouflage and cover of darkness [cf. Tacitus on the Harii (*Germania* p.43) or screening walls of fire (*DBG:* 242-243*)]*. That the Germans could apply forward strategic and tactical planning is demonstrated by the recent excavations at Kalkriese (Schlüter, 1999; http://www.geschichte.uni-osnabrück.de/projekt). The undulating earthwork, topped by a hurdle palisade (*figure 10)*, which created a funnel driving Varus' troops into a prepared killing ground between it and a bog in AD9 took some time to construct. Thus the whole plan - deceiving Varus and persuading him to change his direction of march, and directing that march towards the prepared ground - was a brilliantly achieved strategic ploy of a complexity that is rarely ascribed to 'barbarian' warriors. Generalship is implied strongly here, not only in the plan's conception and execution, but in the evident control of forces, allowing the van of the Roman troops to proceed unheeded while harrying and 'stringing out' the rearguard. Preparation of materiel in advance, and logistics of its supply, are also implied by the employment of red hot clay bullets against the Romans (*de Bello Gallico*: 151).

It is clear that the strategy and tactics of the British annoyed and irritated Julius Caesar very much. The use of stratagems is indicated in the account of the Roman force led into ambush at a place where corn was still standing when all other crops had been cut, and the concealment of stakes below the water in the Thames. The hit-and-run tactics of the chariots and mounted troops obviously made life rather difficult as well - there is a distinct hint of peevishness in the descriptions of the refusal of the British to come to a pitched battle. What is clear is that the British tactics were well practised and reflect a form of battle that was mobile and in which ambush and trick were well known. This implies that warfare in Iron Age Britain had been occurring not just in assault on fortified centres, but across intermediate territories, involving the use of rapid mounted attacks, and ambushes and tricks. This does not sound like formal or ritualised 'champion' battles - where is the glory in concealing stakes below water? Large scale raiding seems to have occurred, met by the development of defensive strategies and mobile protective troops. That such activities may have resulted in the annexation of territory seems to be suggested by accounts of the spread of Catuvellaunian influence, demonstrated, among other evidence, by the distribution of coinage, although a major failing of British troops seems to have been in their inability to combine into larger, perhaps multi-tribal, units. The responses of the Germans and British warriors to Roman troops represent a high level of military organisation, but ultimately, they are of a different order to those of the Roman army itself, being derived from a smaller geographical sphere of activity and a less ambitious scope of political destiny.

Tactics associated with large-scale territorial expansion and conquest

Of the final type of scenario, the large scale territorial expansion and conquest, the Roman parallel offers a great deal of evidence, albeit patchy, sometimes poorly dated and often ambivalent. In this form of warfare the use of professional (generally all male) troops under formal direction is most common. There is standardisation of unit size and composition, equipment and weaponry, and standardised practices on campaign. The use of marching camps, and their succession by forts and fortresses as the hold over the new territory is strengthened, is well attested. Traces of the Antonine campaign in Scotland may be deduced from the identification of small differences in the construction of marching camps, suggesting the routes of particular units. "Châque série identifiable définit une ligne de marche pouvant être attribuée à une campagne précise, pour laquelle on peut parfois suggérer un contexte historique défini, relativement plausible" (Wilson, 1984:67). There is also a group of epigraphic evidence - gravestones, belt buckles with inscribed unit identification, marked leather shoe soles and the like (e.g. examples discussed by van Lith, 1984: 76-80), although much of this seems at best to post-date the actual conflict by some years.

Professional armies are perhaps more likely to engage in planned battles than the more irregular troops of simpler societies. They are also more likely to have a wider range of specialist skills and training appropriate to activities such as sieges. Evidence may be acquired from the sites of large set-piece engagements, such as the siege of Alesia, (St. Jean d'Alise) ,where many features and artefacts have been discovered including ramps, trenches, sapping operations, camp sites and contravallations, accompanied by hidden stake pits, arrows, iron caltraps, human and animal bones and Roman military equipment (Coulston, 2001: 35; Sievers, 1996; Benard, 1996); there is also material from Mauchamp, where a comparison between Caesar's own topographical description, a site and excavated features such as ditches has prompted identification of the battlefield (Reddé, 1996, 35-36) despite the admission and plea that "[il] est vrai que les redoutes romains n'ont pas été retrouvées, mais ces *castella,* nécessairement en terre et en bois, n'ont dû laisser que des trous de poteaux" (ibid: 36).

Similarly flimsy, but nonetheless convincing, evidence of Roman military operations has been discerned at Velsen (Netherlands) where the lead sling shots used by the Roman defenders of this river base in AD28 have been discovered in fan-shape trajectories (ibid:35-6) and at Hod Hill where two round houses were found to have sustained hits from eleven catapult arrowheads. Calculations of the point of origin of these projectiles has been taken to suggest a siege tower constructed outside the hillfort defences to the southeast, and that the object was to target the elite residences within (Richmond, 1968: 32-3, Fig. 14 cited in Coulston, 2001:36). Caesar's dissatisfaction with the British way of fighting seems to imply that his own preferred fighting tactic was the formal, parallel face-to-face battle; this form of conflict is ideally suited to professional armies. The soldiers in these units could be trained, drilled and practised in formations and manoeuvres until perfect; the hierarchical and formal chain of command would be well suited to the direction of 'set-piece' confrontations, as was the case in later centuries with commanders such as Marlborough or Napoleon. The Graeco-Roman traditional development of armies had inevitably evolved this form of warfare in response to an urbanised cultural landscape, where agglomerated centres of population were attacked either by sieges or by formal, even ritual, armed meetings in the relatively unpopulated rural plains surrounding them - a mode of warfare amply described in the Iliad.

Among the correlates of large scale territorial expansion is control or suppression of conquered populations. Caesar's campaigns in Gaul were sometimes accompanied by large scale destruction and punitive actions against the surviving populations, both combatants and non-combatants. Unhappily, the archaeological record offers very little physical support for these aspects of Roman strategy. As Bloemers points out "La tribu des Eburones ... fut totalement anéantie après s'être risquée à défendre son indépendance sous la direction d'Ambiorix. Sur le plan archaéologique, on ne retrouve cependant presque aucun vestige correspondant à ces événements" (Bloemers, 1984:81).

A great deal of the evidence we have of the Roman Conquests of Gaul and Britain is in fact circumstantial - the presence of military style equipment need not equate to war and conquest. Alternative possibilities for the

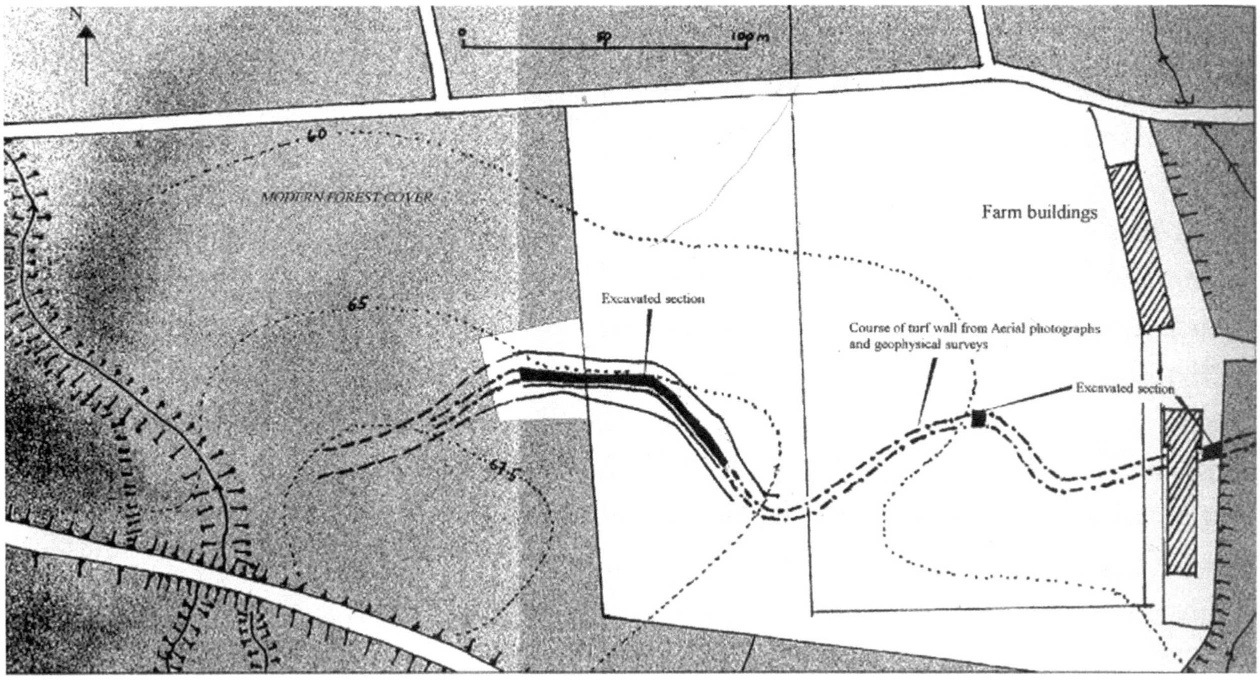

Figure 10: Plan of the prepared earthwork at Kalkreise (Hill & Wileman, 2002:174)

interpretation of this material could perhaps include the presence of merchants' guards, or diplomatic retinues, or the accumulation of souvenirs or booty brought home or traded on by mercenary troops.

The forts and camps, even when they can be dated closely enough to the period, depend for their identification as instruments of war and conquest as much on assumptions derived from the historical sources as upon any hard evidence. In this we recognise a side effect of having an 'historical' source - we enter the realm of self-fulfilling prophecies - the evidence dates from the period of the events described, therefore the events described are 'proved' by the evidence, and thus the evidence is assumed to be from the events described *ad infinitum*.

A similar lack of information recovered by archaeological means has been noted by Hamilton (1995) for the Trier region. There, the Roman Conquest is only archaeologically visible, in many cases, through a longer term change in material culture, architectural style and settlement distribution.

Conclusion

In conclusion, it appears to be possible to recognise particular tactics, and to some extent larger strategic organisation, in the archaeological record for prehistoric war, within limits imposed by evidence survival and understanding of the range of options available to contemporary societies such as weapons technology, social structure and political complexity. However, like other forms of evidence previously discussed above, the evidence is at best circumstantial, incomplete, and liable to a variety of interpretations. Once again, it depends on *context* and on the ability to *confirm* the evidence through other means - history, evidence of cause, evidence of occurrence and evidence of results. A discussion of the kind of evidence associated with the consequences of war is discussed in the next chapter.

Chapter Four - The Effects of War

The previous discussions lead to a recognition, albeit hedged about with many doubts and alternatives, that warfare has certain correlates, and the onset of war has a variety of results. The correlates of war include features such as defensive siting or building, increased rapidity in the development and technology of weapons and armour, increasing social complexity, social and environmental stress, and political change. The results of warfare in small agricultural prehistoric communities are mainly a matter for conjecture, but this can be based on examples, both modern and ancient, and upon commonsense extrapolation. Results of aggression will differ according to the political, economic and technological complexity of the societies involved, and in relation to causal factors, aims of attack or defence, and duration of hostilities. In terms of northern European prehistory, the most likely forms of warfare seem to be those typified by raiding, status and endemic war, with isolated occasions of expansive territorial warfare. More developed forms of professional warfare do not seem to occur until the first century BC and the advent of Roman expansion in the European sphere.

War effects can be categorised as short term, medium term and long term. In the short term, there are the bodies of the defeated, the smoking ruins of their houses, villages and crops, the refugees created by the destruction and the relatively sudden cessation and abandonment of previous forms of settlement and subsistence. Medium term results are those related to resettlement, clearance, burial and seizure, while long term results may include realignments of trade and political alliance, cultural change, absorption and dominance of other cultural groupings and changes in land use and production.

Short term effects of war

It is notoriously difficult in archaeology to identify short term change - our ability to date events or to recognise ephemeral effects remains too imprecise. At the same time, while we can readily identify long term change, we find it very difficult to isolate the causal variables of that change amongst the myriad of possible factors that can be imagined. The problem is exacerbated by the fact that too much archaeological research in the past has been site-specific, and not enough attention has been paid to the search for regional and territorial patterning of change events. The exception to this is the interest in climatic change and its effect upon agricultural regimes that has been developed by many recent researchers, but even here a macro event is postulated, without much consideration of the accompanying social and political activities that may have been instigated by the incremental, and perhaps generally imperceptible, alteration in conditions. It is easy enough to talk about climatic deterioration over a period of a couple of hundred years; it is much harder to imagine how such a decline may have been understood and experienced over eight or nine generations. Haselgrove (2002:55) has noted these problems in his discussion on changes in Belgic Gaul resulting from the Roman invasion; he has also stressed the problems created by the relative 'crudeness of the archaeological timescale' and the unreliability of much previous data collected without modern rigour.

The fall in yields, the increase in stock and crop disease, the change in trading patterns, the frequency of famine periods and the like resulting from climate change would become matters of family legend. Each succeeding generation would have had to make their own assessment of grandfather's stories about how much better things were in his day. The point at which the environmental stress becomes too great for inaction is hard to gauge. It is probable that some other catalyst would have been involved - the reaction of neighbouring communities, a particular disaster, an outbreak of sickness, a forceful individual - all might turn the tide of endurance into action for change. That change itself may take many forms, depending on contemporary choice availability. If there is unoccupied better land, then a peaceful move may be perfectly feasible; if the better land has been abandoned due to the deaths through illness of its inhabitants, then again peaceful shifting of settlements is possible. Only if a number of communities are vying with each other for alternative resources are the conditions right for aggressive action, whether instigated by large scale necessity or the action of a forceful individual using perceived necessities as a rallying point. Once this process begins, it can easily escalate; in nineteenth century German West Africa, "Every small chief who had a few cattle plundered and murdered the others who had rather more, and everyone who had been despoiled looked for a third at whose expense he could recoup his loss. Bloodshed and misery, murder and horror, were so prevalent throughout the land..." (Vedder, 1966:251).

If sudden aggressive action has taken place, there are a number of features we might expect. We would find evidence of weapons trauma on bodies, as already discussed; in addition, there should be traces of trauma in the landscape. We might find destruction horizons in settlements. Crops might be abandoned or destroyed in the fields; storehouses may be looted or burned. Evidence of sudden abandonment - the leaving behind of expensive but bulky or heavy equipment, perhaps, or animals left to die in pens and stalls - might be a strong indicator of unplanned flight before an enemy incursion. Trees may be chopped down, ritual centres desecrated, monuments defaced.

Beginning with the human cost of war, we can note that the experience is not just of the mortality inflicted by battle, but the far greater effects of the loss of a generation of young adults, the loss of a work force, and the effects of displacement - famine, disease, stress and despair - that will have affected populations. In extreme cases, a whole race can be doomed to extinction - as occurred in Tasmania in the nineteenth century. Similar abuses took place in New Zealand, reducing the Maori population from about 240,000 before colonisation to just 40,000 by 1896. Near extinction episodes occurred in the Americas - estimates of the pre-colonisation population of Mexico suggest some 12 million people, which by 1600 had fallen to just over one million; for Peru, the figures are nine million falling by 1620 to 600,000 (Cotter, 1999:111). Several Native American tribes in the United States did become fully extinct. To a large extent, it was the introduction of Old World diseases, to which the native peoples had no resistance, that destroyed these populations - smallpox, typhus, influenza, measles, mumps and whooping cough. However, their debilitation due to the loss of subsistence and to abuse was a major factor.

Estimates of the effect of the Roman conquest of Gaul after Alesia in terms of population decline range from 14% to 28%; "On aurant tort de considérer ces pertes comme négligeables, elles durent provoquer pour un temps dans certaines régions, un dénatalité et un ralentissement de l'économie" (Deyber, 1996:90). In the vicinity of this campaign,

> *"La terre n'en souffrit pas mains alentour: tout l'arrière-pays avait été mis en coupe réglée, par les Gaulois de l'armée de campagne de Vercingétorix d'abord, par les Romaines ensuite, mais encore par l'armée de secours par suite de l'absence de logistique approprié. Il n'est pas possible de mesurer actuellement l'ampleur de ces dévastations, que l'archéologie seule pourra permettre de préciser ultérieurement"*
>
> (ibid:91).

Illness, too, is devastating. During the occupation of Greece in 1941, the population suffered plagues of rats, famine, and outbreaks of infectious diseases especially tuberculosis, with hundreds dying every day from hunger and related disease (Glenny, 1999:482). A stressed prehistoric population might expect similar trauma, and though it might necessarily have been on a smaller scale, the proportion of the population affected might have been as great, if not greater. The combined effects of the loss of the active work force, trauma, destruction of homes and means of livelihood, and the burden of caring for the bereaved, the wounded, the orphaned and the elderly or sick deprived of their usual supports, can be totally demoralising and exhausting for survivors.

At the same time, like the Incas or the Moslems of Bosnia, they will have seen the desecration and destruction of their cultural symbols and religious centres; mental illness is a likely outcome for many, and certainly most people have to deal with severe distress. Destruction of housing stock and displacement of people are major features of the short term results of warfare. In archaeology, we constantly see abandoned settlements but rarely ask where the occupants went, and what happened to them. It is very rare indeed that we can say that the people from this village got up and rebuilt at another identified site. It therefore seems essential that the processes and forms of abandonment and settlement need to be studied in much greater detail over wider areas.

In Croatia, a quarter of the housing stock was destroyed or severely damaged during the recent wars, about a quarter of a million refugees left the area and a slightly greater number moved in from neighbouring Bosnia and Herzegovina (Gasparovic et al, 1995). It has been estimated that at least 60% of all homes in Bosnia were damaged or destroyed (Amanpour, 1996:1). A 15 per cent decline in population has been reported from Herzegovina, with the population dropping from 4.4 million in 1991 to 3.7 million in 1997, and about 25% of the population of Bosnia has been lost to displacement (OHR, 1998). These ratios are not dissimilar to those estimated for Gaul in the aftermath of the Roman conquest.

The psychological effects of warfare may not easily be appreciated archaeologically. The effects of trauma on populations, however, cannot be under-estimated. Referring to earlier troubles in the Balkans, it has been suggested that

> *"Fear is the dominant, the ever-present motive. It builds villages. It dictates migrations. It explains deceits. It has created the morals of a country. The Bulgarians are, of all races, the most solid and enduring... It is no common shock which wrings a cry from them ..and yet fear is the great fact of their daily lives ..a woman who had barked like a dog since the day her village was burned; a maiden, who had become an imbecile because her mother buried her in a hole under the floor to save her from the soldiers; a lad who turned ill with 'fear' from the moment when a soldier put a knife to his throat; children who flee in terror at the sight of a stranger... "*
>
> (Brailsford, 1906:36-7).

Livestock are also frequently casualties of war. Animals may be abandoned in their stalls or fields to die; they may be killed to feed the incoming victors in the short term, or driven away with the refugees, they may also die of stress or be slaughtered to feed people bereft of their crops and stores. After the Second World War, parts of Yugoslavia were recorded as having lost much of their stock. In the Cazin District, "56% of the agricultural inventory had been destroyed, 20.7% of all houses .... The number of horses had fallen by 60%, sheep by 63.2%, cattle by 55.6%, swine by 58.7%. The herds were exhausted, undernourished, ill" (Krzisnik-Bukic, 1991, 3). After the more recent conflicts in the former Yugoslavia, a similar picture of devastation of livestock has emerged, as has been previously mentioned. The Lika region had seen an overall livestock reduction of 50%, as people sold their cows cheaply or were forced to slaughter them for food. Overall, 120,431 cows and pregnant heifers were either killed or misappropriated in 44 occupied districts, together with 65,280 sows and pregnant yearlings, 8,042 mares and pregnant yearling mares, 231,000 breeding sheep and 2,648,000 poultry (Richardson, 1995). Studies of livestock populations in prehistory may, therefore, be able to add to the body of evidence for traumatic events such as war.

Displaced populations might be discerned in the destruction of natural habitats - woods cut down for fuel and the building of temporary shelters, and wild animal populations decimated for food, even species not usually hunted, such as the mountain gorillas of Rwanda. Exotic species may be overhunted for their trade value as well as for meat. The people doing the killing may not be skilled hunters, and do not necessarily appreciate the value of preserving breeding stocks or avoiding female animals with young to rear; indiscriminate slaughter may therefore devastate local wild populations.

Disturbance and intimidation (especially of mammals and birds), plant destruction, contamination, and habitat destruction will lead to the extinction of some animal populations, especially the most threatened and population reductions in those species which can be sensitive to disturbance, and even the complete loss of natural habitats in many locations, in which the fauna of such regions will be transformed. In Azerbaijan, conflict has resulted in 1.1 million displaced persons; among other effects has been massive deforestation as people living in temporary camps have sought fuel in the cruel winters. It is considered likely that this will over time lead to considerable erosion (Turvalev & Hajiyev, 1994). Several endangered species have been identified as being in great peril, including the Caucasian bear, and deer and wild boar have been hunted to the point of local extinction for food.

Many of the Armenians affected in Azerbaijan had been pastoralists, practising transhumance over nearly 245,000 ha. of pasture; access to these pastures has now been lost, as has access to important fresh water springs for their remaining stock. They have also been cut off from the mineral resources that represented work and wealth for many other communities. More than 645,000 ha of cotton, vineyards, wheat, corn, and vegetables have also been abandoned (Turvalev & Hajiyev, 1994). These effects may have been similar in kind, if not in scale, to those experienced by prehistoric survivors.

Situations like these not only create much suffering for survivors of the original conflict, they lay the foundations for future outbreaks of hostilities and terrorism - "...when interwoven with such elements as population growth, poverty and injustice [environmental problems] increasingly contribute to existing and future political tensions and their escalation to violent conflicts" (ENCOP, accessed 2002).

Medium Term Results of War

In the medium term, perhaps a year or more after actual hostilities, there may be evidence that cropland has reverted to scrub, ditches and drainage systems have silted up, and temporary camps have achieved a greater degree of permanence. The first signs of alien colonisation may become evident; land use and communication routes may be in the process of changing. Rebuilding may begin, perhaps in a less well constructed or different form; new structures may present a different spatial morphology to those destroyed.

The destruction of local or national infrastructures during a war result in a longer term process of financial hardship and loss of production. No figures are available for these types of effect in prehistory, but we can use (with reservations) a more recent conflict to suggest something about proportions of war effects, and consider a variety of evidence for change around the time of Roman expansion in Northern Europe. The war in Croatia resulted in damage or destruction to 30% of the country's industry, 9.3% of its agriculture and processing capabilities, 8.5% of its forestry and wood processing, 12.3% of its tourism and catering and 16% of its traffic and communications installations (Gasparovic et al, 1995:157). More than that, "The influence of the war has changed the social structure......due to the war and insecurity of life, the total population and particularly displaced persons, the rate of national growth of the population (which was already very low in Croatia, according to some estimates) has become negative, i.e. the number of new born is now lower than the mortality rate" (ibid). This effect may only last for a couple of generations, however, and be hard to identify archaeologically. It also seems to be the case that wars are often succeeded by 'baby-booms'.

In Bosnia and Herzegovina, 75 per cent of refugees (550,000 to 610,000) were still abroad and only 190,000 refugees had returned by.1998; 85 per cent of displaced persons (950,000) were still displaced and overwhelming ethnic majorities had been formed in all parts of Bosnia and Herzegovina: the highest minority ratios are about 13 per cent (in the Tuzla-Podrinje and Sarajevo Cantons) (OHR, 1998). Thus the demographic profile of most regions has been ethnically altered to a considerable degree. People have moved away from war damaged areas, and areas close to frontlines of the conflict. There has been change, too, in the way people live. Before the war a relatively large number of families shared their homes with relatives; now households have split up. There has been a speeding up of the drift to urban centres which had begun before the war, and reluctance, especially of the young, to return to villages once they have experienced urban life. "People have not gone back to areas where there are little economic prospects. Regardless of agreements they had made, many families have not returned to remote villages (even after their houses have been repaired with international assistance)" (OHR, 1998:7). These data are not strictly relevant to prehistoric situations, but they highlight the kind of demographic changes that may result from conflict, and may need to be considered.

Dislocations of settlement patterns are relatively common in the archaeological record; for example, Haselgrove's study of Belgic Gaul noted that the period from about 70 to 25BC was one of widespread change, with a high proportion of Iron Age sites being abandoned, and most Gallo-Roman sites being founded.

> *"The evidence such as we have suggests considerable instability in the first century BC settlement pattern, evinced both by the short-lived fortified nucleations and by the rural sites, some of which were abandoned for good, some reoccupied after a short break and still others, completely new foundations...Two of these certainly...look like implanted settlements, and the lack of any indications to the contrary, such as evidence of more gradual growth, is striking. Overall, a 'crisis' model ... of settlement centralisation during a period of sustained military or socio-economic stress, offers a far better fit to this data, than the more conventional model of an evolving settlement hierarchy. This view would link the foundation of the fortified settlements such as Villeneuve St. Germain to the abandonment of many of the rural settlements during a period of sustained crisis, the population subsequently dispersing back to the land when this was over, or moving to other nucleated settlements that replaced them"*

(Haselgrove, 2002:66).

It is surely tempting to view this evidence in a similar light to that of the former Yugoslavian states.

Bloemers, studying Lower Germany for the early Roman period, has found evidence for the foundation of an entirely new type of settlement with non-native inhabitants and a very restricted economic interaction between settlement and native society in the Nijmegen area. His findings are summarised in the following list:

> *"1. At all sites, indications are found for the development of a large and probable civil settlement, where intensive building activities started about AD40 or at least during the reign of Claudius.*
> *2. All sites had some sort of military presence of Augustan and sometimes Tiberian date near to the civil settlement. This military presence seems to have stopped before, or at least during, the early phase of the development of the civil settlement.*
> *3. No site has yielded any clear indications for the presence of a native pre-Roman predecessor, and consequently, the location is discordant with native settlement patterns.*
> *4. All sites have dimensions much larger than the known permanent native settlements in the region; have some sort of regular layout and have, as far as is known, buildings which differ in construction and plan from the known native houses.*
> *5. All sites have large quantities of Roman and Gallo-Roman products.*
> *6. As far as expressed by Roman imports, contacts with the rural areas around these settlements seem to have been very restricted before the middle of the 1st century AD. They increase strongly from then on"*

(Bloemers, 2002:83-4).

A contrast may be drawn with the conclusions of Elizabeth Hamilton (1995) with regard to the Trier region in the same period. Her analysis of the evidence from the period 100BC to AD150-200 is that "...there is nothing in the material record, as opposed to the written record, that would urge upon the observer the conclusion that a military invasion and occupation had taken place here. What most modern archaeologists would conclude is that a relatively slow (150-200 years) process of diffusion of artifacts and, later, some ideas and habits, had occurred" (1995:42). In other words, despite the written evidence for a cataclysmic eruption of violence and destruction, the evidence offers little or nothing that could differentiate events in this region from those in essentially quisling East Sussex in the first century AD (Rudling, 2003: pers.comm.), or from cultural diffusion of a consensual kind. Even the appearance of Roman style buildings means little and what little military equipment there is might be explained away as a diplomatic presence. The narrative component of our understanding of the period tends to demolish the clinical methodology upon which we pride ourselves when dealing with prehistory. Both become suspect when subjected to comparison of this nature.

Longer term effects of war

In the longer term - from several years to a generation after the event - the cultural package of a region may be significantly altered by importation of new ideas, styles and products of the victorious group. Communication and trade routes may have become realigned to reflect the change in political and economic history and interests of incoming communities; material culture styles may have altered to reflect those of the conquerors, and different forms of ritual behaviour may be apparent. Land use may change, and the status and wealth of the conquered may have stabilised at a distinct level from that of the victors.

It is sometimes possible to trace very long term change - changes which persist for hundreds of years. In Guatemala, the Mayans still make up between 50-80% of the population, still use their ancient languages, costume and ways of worship, and follow their ancient calendar. But they are a subordinate people. Their life expectancy is an average of 17 years lower than that of people of mixed or Spanish descent, in 1987 it was reported that two thirds were living in extreme poverty, and a 1979 report showed that 98% of them owned no land or insufficient land upon which to support their families. Efforts to protest at their conditions have lead to extreme reaction from military governments and semi-official death squads; since 1980 50,000 Mayans have been killed and one million displaced from their homes (Cotter, 1999:367).

In the United States, unemployment, poverty, suicide, crime and alcoholism rates remain far higher among Native Americans than among other communities. The situation in South Africa remains difficult for most blacks, with the majority of wealth still remaining in white or Asian hands. Admittedly, these examples relate to imperial wars of conquest by people of very different ethnicity and technological development to that of the conquered population; the differences between communities remain very evident in terms of skin colour and physical characteristics, making assimilation more difficult than might be the case if both sides had similar origins. Nevertheless, it is worth remembering that tales of 'Saxon' opposition to the Normans had a long currency in the medieval period, and became enshrined in the popular imagination, where they remain to this day. A conquered population is likely to have a long memory, and may even have a tendency to dwell on past glories and disasters rather than to deal with present and future problems.

The evidence for agriculture and industry in post-war situations shows that despite international aid, recovery is often slow; there are often fundamental changes in the nature and products of trade. In Croatia, the agricultural population has seen a decrease from 43% of the country's total in 1961 to 15.2% in 1981, partly due to global trends towards urbanisation and partly due to the war (Richardson, 1995). In 1990 family farms produced 50% of the country's wheat, 76% of its corn, 70% of the meat , and over 92% of its milk, but a quarter of them also needed an outside income to survive. The production of sufficient food, therefore, is dependent on the return of displaced persons to these family farms - the very process which has not occurred - or massive investment in the state-run farms along with foreign imports, investment which is likely to lead to long term national debt and stagnation. It is not simply a question of resuming production - the infrastructure must be replaced, the workforce must be found from a generation that has lost many of its young men and women, and the lure of urban life must be compensated for by better services and housing in rural areas.

Detailed regional studies may be able to discern changes in settlement and production in prehistory that could be ascribed to such disruption and changes in attitude, although the causality is likely to be hard to establish. Nevertheless, abandonment of farms and settlements in marginal or frontier areas may not be always due to soil degradation or non-combative insecurity. The problem lies in recognising whether such abandonments might occur *before, after* or *instead of* periods of aggression.

It is rarely that communication routes are recognised is a weapon of war with longer term results; the situation in Israel may, however, offer parallels to be considered when studying earlier conflicts, particularly, perhaps, the Roman conquest of Gaul and Britain. In the occupied West Bank, roads have been built by Israel to connect their settlements with Israel itself. These wide, straight roads bypass the Palestinian settlements and refugee camps, marginalizing the Palestinian people. In this conflict,

> *"Mobility became a commodity and a source of political control and economic value"*
>
> (Selwyn, 2001:234).

It is not suggested that the Romans operated in the same way; it is possible, however, that their roads, built to

	COLONIALISM	IMPERIALISM
ERADICATION – RESETTLEMENT	Abrupt cultural change (resettlement)	Regional 'empty cell'
ACCULTURATION	Slow indigenous culture change	Slow indigenous change in economics
EQUILIBRIUM	Settlement enclaves	Indigenous cultural

Table 1: A matrix of probable behavioural outcomes in situations of power domination, based upon ethnographic and historical examples: (After Bloemers, 2002:fig.6)

facilitate control of provinces after the Conquest, and to service imperial official among whom would be numbered tax collectors and government officers, would be seen as a symbol of control by force.

Attempts have been made to characterise the differences between colonial and imperial expansion and strategies of invasion by one culture of another. One such model is that offered by Bloemers (2002 - Table 1), in which cultural change is mapped in relation to possible scenarios.

This may be a useful outline for larger scales of prehistoric war, although over simplified, in that it seems to ignore the contribution to the record of resistance by the indigenous culture, or different reactions by different sectors of the indigenous society (such as aristocratic compliance but peasant resistance). Even on smaller scales, something of these processes may have occurred as a result of war where local identities have been developed; local and kinship connections for subsistence, trade and communications may have been similarly disrupted and changed by movement in of other groups either peacefully or with hostile intent; in 'colonial' situations, it may be possible to note contemporary but different social groups operating at comparable levels, while in 'imperial' incursions, the indigenous culture may become visibly absent in some places.

One of the issues raised by this model is that of cultural assemblage; invasion and conquest might normally be expected to be accompanied by new objects and styles, and new trading patterns. These may not be immediately highly visible. It may take a long time for the new assemblages to filter down into poorer levels of the remaining indigenous society, and tribute, tax and extortion could combine to delay their potential for acquisition among the conquered. It is also possible that only part of a new cultural assemblage becomes common in a conquered territory, either because control of exotic or status goods remains in the hands of elites, or because of a conscious decision on the part of the victors to deny certain goods to subject peoples, or because of rejection of parts of the package by the conquered population as a form of resistance. It is probable, in fact, that all these mechanisms could operate at the same time. Thus adoption of new styles and customs would appear patchy in the archaeological record, very much as Hamilton (1996) describes for the Trier region in the 1st century BC.

Conclusion

These effects of war can be recognised as highly complex, mutative and difficult to assess in terms of the archaeological record. Nevertheless, wars do have effects, and if warfare did occur in prehistory, it is clear that some traces of these effects should be recoverable. The interpretation of their cause is the challenge; most effects could have had multiple causes, but it should be remembered that warfare is one such, and its possibility should not be ignored.

Consideration of effects of modern wars may highlight forms of behavioural complexity not otherwise associated with prehistoric societies, although extreme care must be taken not to impose modern assumptions on the evidence. Modern examples are included here in order to stimulate discussion rather than to draw interpretative parallels.

Part Two - The Identification of War

Chapter Five - Correlates of War

The various discussions and evidence assessed in previous chapters may lead to the identification of an extended list of correlates related to the causes, occurrence and results of warfare, which include traditional forms of evidence but which add further dimensions to the recognition of the processes of war as they might have applied in prehistory.

These correlates might lend themselves to the creation of a form of check list for warfare, against which data from any one site, group of sites or region could be matched, providing a foundation for a more informed and complete set of variables. Given the proportion of present variables identified in each case, an assessment of the likelihood for the occurrence of warfare could be made with a little more certainty than has previously been the case, although in no way could such an exercise constitute *proof* of warfare or its absence. The existence of an extended set of correlates could also be used to inform research designs to add weight to interpretations of social or cultural change that might have been (or might not have been) associated with conflict.

It is always important, however, to recognise the dangers inherent in the creation of models as meta-explanations; despite this, it is often necessary to use such aids in any attempt to recognise patterns of human behaviour in prehistory, whilst remaining aware of the dangers of self-fulfilling prophecies and universalising assumptions.

Each society, locality and period is likely to have its own set of particular correlates; what this study aims to do is to identify those correlates which have generalised and broad applications across periods, cultures and geographies - a macro checklist. They are divided into those correlates which might be associated with the causes of war, preparations of societies faced with hostile potentialities, evidence associated with destructive functions of warfare, and factors which might result from the occurrence of war.

The present exercise must be approached with caution and the acceptance that correlates are only useful in *combination* and in *context*, and that interpretations based upon their presence or absence are only ever conditional.

Some of these correlates have already been discussed, forming as they do part of the existing and much cited forms of evidence for warfare. Others need to be considered a little more carefully, because of their interpretative ambivalence or their particular emphases. It is not considered necessary to rehearse all the headings in great detail - for many, a large literature is already in existence which it would be redundant to repeat here. Instead, focus will be placed on those which may not have been considered in great detail by other commentators on the topic.

Correlates potentially associated with the process of war in prehistory

Causal

i) Climatic deterioration over the period
ii) Demographic rise
iii) Presence of other subsistence/social threats
iv) Increase in social complexity
v) Territoriality developing
vi) Changes to access for resources

Preparatory

vii) Appearance of fortifications
viii) Proliferation of weaponry
ix) Iconography possibly related to warfare
x) Abandonment of settlements
xi) Appearance of defensible settlement strategies
xii) Changes in settlement location
xiii) Changes in settlement morphology
xiv) Appearance of guard settlements
xv) Settlement locations related to line-of-sight links
xvi) Change in subsistence patterns

xvii) Change in trading patterns
xviii) Unoccupied zones developing

Functional

xix) Burned sites
xx) Unburied bodies/weapons trauma (human and animal)
xxi) Weapons damage

Resultative

xxii) Change in subsistence patterns
xxiii) Change in trading patterns
xxiv) Trophy collection and memorials
xxv) Iconography possibly related to warfare
xxvi) Deviation in sex ratio
xxvii) Depopulation
xxviii) Unoccupied zones developing
xxix) Population movement
xxx) Abandonment of developed arable
xxxi) Changes in settlement location
xxxii) Changes in settlement morphology
xxxiii) Changes in biodiversity due to human relocation
xxxiv) Changes in health profiles/standards of living
xxxv) Artefactual and cultural change
xxxvi) Changes in ritual and burial practices
xxxvii) Territoriality developing

Causal Correlates

i) Climatic deterioration over the period

Climatic deterioration is recognised as a catalyst for social and cultural change, because of its tendency to create conditions of subsistence stress or fear of such stress. This topic has been discussed in Chapter One. A possible source of evidence for conflict in prehistory may be changes in the siting and interrelationships of settlements and their associated fields and communications. A number of studies carried out in the Southwestern United States and Northern Mexico have looked at these features and their interpretations as adopted by cultures within or neighbouring several Puebloan societies in the period 700 to 1400AD. It may be useful to compare some of these studies and their results with the situation that presents itself in southern central England in the Late Bronze Age/Early Iron Age. At first glance, this seems an unlikely comparison, but the two regions and periods do, in fact, share a number of parallels. In both instances, the technical level of the societies was broadly similar, although far less metal was worked or used in the American societies. The primary forms of settlement were isolated farmsteads or small villages at the beginning of the period, with settlement aggregation or an increase in settlement density with concurrent abandonments, and the eventual building of 'hillforts' in both regions. There are broad variations in the size of Pueblo groups, and in Late Bronze Age British settlement regions, and in some cases these are not dissimilar (for figures see the Case Studies in Part Two). Subsistence agriculture was practised in both cases utilising small fields and gardens, with some stock farming or ranching. In both cases, there are typical forms of evidence for warfare present, and in both cases, these have been called into question at various times by interpretations that would prefer to see change as unrelated to aggression.

Both cases were likely to have suffered the effects of one of the identified causes of war - climatic change. The reorganisation of the Southwestern cultures occurred during the period of development and impact of the climatic downturn known as the "Little Ice Age", an event which eventually affected Medieval Europe. Late Bronze Age English societies were also existing in a period of climatic deterioration, although there is still some debate over the cause.

Climatic deterioration in the Southwestern United States between the 8th and 15th centuries AD resulted in shifts in rainfall patterns leading to increased desertification, extending the margins of the Sonora desert and creating an increased demand on the resources of the river drainages. Studies have identified major periods of environmental degradation between 750-925 AD and 1275-1450 AD; during the later period the "spatiotemporal pattern in annual rainfall was significantly disrupted" (Adler et al, 1996:421), resulting in chaotic variation in seasonal precipitation. Higher altitude regions would have experienced drier and colder conditions, with shorter growing seasons and increased risk of frost damage to growing crops. There is some evidence to suggest that the option of an increase in stock farming was also adopted in parts of this region as a result.

Evidence for the Bronze Age climatic event is discussed in more detail in Case Study One; some studies suggest a major climatic downturn around 1000BC, based on evidence from dendrochronological studies (Franzén and Larsson, 1998) who also cite evidence from texts, environmental data, glacier expansion and low tree limits, and records of violent increases in water levels in rivers and lakes. Other studies have suggested that the Late Bronze Age saw the beginning of an unusual rise in precipitation (Lamb et al, 1966) and temperature changes resulting in an overall decline of crop ripening levels (Parry, 1978) and increased peat growth (Barber, 1982). This, it is argued, caused general contraction of settlement, abandonment of marginal lands, general retreat from the uplands, and far-reaching dislocation of settlement and agricultural systems. Indications of new crops, more intensive farming methods and food collection and storage methods, with possibly severe population decline are suggested by Burgess (1985). If this deterioration occurred, farming strategies would have been forced to adapt to different conditions, involving abandonment of higher arable fields and an increasing reliance on stock farming. Settlement might be expected to relocate to sheltered terraces above seasonal floods. The Middle Thames region may have experienced effects including longer, colder winters, more rain, localised flooding, cooler summers and a curtailed growing season; in the Thames Valley bottom and terraces, dairying may have provided an alternative to cereal growing.

As in Britain, there has been much argument in the United States about whether the changes in settlement and subsistence patterns observed during this climatic change owed anything to a concurrent or subsequent increase in hostility between groups and polities. Alternative explanations include change in social stratification and hierarchy, variant responses to environmental exploitation needs related to climate stress and/or increasing population density, and social and cultural changes and movements. In Britain, explanations for change in the Later Bronze Age include a similar list, with a number of possible ritual factors added to explain some features.

ii) Demographic rise

A rise in population number is also recognised as a potential source of subsistence and social stress, and again, this topic has been discussed in Chapter One.

iii) Presence of other subsistence/social threats

There are many other, possibly intangible, levels of threat which may have been perceived by prehistoric societies, which might have precipitated conflicts. Obvious examples might include geomorphological change (rise or fall in sea levels, volcanic activity, river course change and so on), anthropogenic effects (soil exhaustion, deforestation, effects of canalisation) and cultural effects (effects of religious or ritual practice, inheritance practices affecting viability of production systems, sex ratio manipulation). In some cases, these may be identifiable through the archaeological record in particular or vulnerable localities; quantification of these threats and their effects would probably be restricted to results for individual regions or cultures.

iv) Increase in social complexity

Political factors are clearly implicated in the causes of war; again, this has been discussed earlier.

v) Territoriality developing

Agricultural development and intensification create the need for the construction of permanent boundaries in the landscape - to protect crops from stock or wild beasts, and to protect stock from predation and straying (Yates, 2001:66). It has been suggested that as societies put greater investment into their landscapes, the recognition of those bounded spaces as property, and therefore as *owned* places, grows (Field, 2001:60; Thomas, 1998:8-9).

Once this perception of the landscape is in place, new distinctions emerge - those who belong, and those who do not. At the same time, the boundaries of a demarcated landscape act to constrain expansion and to limit reactions to change of those living within it. This developing territoriality can lead to aggressive responses for a number of reasons: problems of trespass or raiding by outsiders, problems of competition for desired land by neighbouring groups, the desire to increase resources for insiders through expansion of territory, and the limitations of the enclosed space when subjected to stresses such as population increase, climate change, crop, stock or human disease, or political change. Once the landscape has been demarcated by permanent boundaries, the choices available to those living within it are greatly reduced, and friction is more likely.

vi) Change to access for resources

If natural or political change results in the closing of channels of resource, stress is likely to result. Pre-existing relationships of wealth and/or status may be disrupted, creating conditions which may allow the rise of intra-group or inter-group competition which could lead to hostilities. Lack of essential resources could lead to a response from a whole community, such as migration or war, and lack of status resources may encourage raiding (Helbling, 2006). Foreign trade intervention may also encourage raiding behaviour, as seem to have been the case in Gaul before the Caesarian conquest.

Preparatory Correlates

vii)Appearance of fortifications

Decisions to provide fortifications strongly suggest a perceived need for defence, and therefore a perception of threat. The problems associated with the identification of fortifications and their purposes have been discussed in Chapter Two.

viii) Proliferation of weaponry

Any increase in numbers and types of weapons in a particular culture may signify an increased preparedness for conflict; the topic has been discussed in Chapter Two.

ix) Iconography possibly related to warfare

The general topic of iconographic evidence has been discussed in Chapter Two; iconography as a preparatory correlate might relate to the signification of identity, of threat and of warning, and to the demonisation of enemies - in modern parlance, propaganda. Identification of such material in prehistory is likely to be virtually impossible.

x) Abandonment of settlements

Settlements and houses excavated by archaeologists have all, *de facto*, become abandoned. There are many reasons why people choose to move away from a site, including changing climates, personal circumstances, soil depletion, disease, changes in economic strategy and lots more. Warfare is, therefore, hard to prove as a reason for abandonment. However, by studying the processes and evidence of different types of site desertion, it may be possible to establish which sites were not likely to have been abandoned because of hostilities, and those where warfare as a cause for leaving is a possibility. Settlements and landscapes are also changed by warfare, and similar studies may be useful in assessing the causes for variation in surviving communities and subsistence.

Factors which affect the interpretation of the causes and motivations behind site abandonment and their relationship to warfare include the speed of abandonment, whether or not there was any anticipation of return - such as in the form of artefacts cached or prepared for storage, whether abandonment is likely to be seasonal or periodic under normal circumstances, whether abandonment is restricted to certain areas within a settlement, or whether abandonment affects not just one settlement but several or many within a region which would not otherwise be likely to be deserted wholesale at one time. Dating and the assessment of contemporaneity present the biggest problems for current research into these issues. It is not generally possible to be able to fine tune dating evidence sufficiently to prove similar actions within a couple of generations, let alone in response to a single event.

It is therefore tempting not to try. Longer term factors such as agricultural impoverishment are easier to assess, and indeed are probably implicated in a great many abandonments; the stress of crop failures, however, may also be a contributory factor to conflict, and the two explanations need not be mutually exclusive.

The way in which a site has been abandoned has been the focus of also be a contributory factor to conflict, and the two several studies. Schlanger & Wilhusen (1993:90) have created a table which illustrates some of the main features of different types of abandonment, a modified version of which offers the following guidelines for their recognition (see Table 2 below).

"When structures are abandoned, people are forced to make decisions about what to take and what to leave, and how to adjust the household tool inventory to changes in structure function" (ibid: 91). This fairly obvious observation nevertheless begs a number of questions for archaeological assessments. It has been suggested that artefact weight is a useful dimension for predicting discard behaviour, depending on assessment of remnant use-life, replacement cost and portability (ibid:91-2); the highest weights tend to be associated with burned dwellings, and the lowest with those left intact (salvage having occurred). In planned abandonments, certain items of tools and equipment may not be taken away to the new house or settlement. "The analysis suggests that artefacts diagnostic of site function, as well as those diagnostic of site use (e.g. food preparation, tool manufacture), may be left at permanently abandoned sites if they have low overall replacement costs (e.g. artefacts that have low raw material replacement and manufacture costs) or are worn (exhausted) or broken non-functional items" (Tomka, 1993: 23). It is not unusual for material to be left at abandoned sites for a number of other reasons - the intention to come back and pick them up later, the possibility of more permanent return at a future date, or the decision to use the old site for another purpose - as a field barn, for example. This results in "delayed curation" of implements cached in expectation of return; where the return is delayed or increasingly spasmodic, caches are reduced. There may be occasional visits to pick up this and that; and improvised or worn objects tend to be left longest or eventually to be discarded at the abandoned site.

People being forced to flee in a hurry are theoretically likely to take with them those items which either have value for immediate survival or intrinsic value for trade. They might take food, clothing, blankets, basic kitchen equipment, tool kits, water containers, personal kit, and small items of family memorabilia or sentimental meaning, as well as (especially in modern times) necessary documents and records. They may attempt to take some livestock - a horse or two, perhaps a cow and a basket of chickens, and probably the family dog. They are not likely to attempt to round up stock from the fields, or to bring heavy or bulky goods such as ploughs or quernstones. Theory, however, seems to go out of the window in real life situations. Countless photographs and film images of refugees during the twentieth century have shown families laden down with all manner of furniture and goods, trudging along roads to nowhere. In 1810, Wellington's army had great difficulty moving back to their prepared positions on the lines of Torres Vedras because of the fleeing Portuguese blocking the roads, trying to drive pigs, hens and cattle with them, wagons overturning, horses dying, and spilt and broken cases and boxes littering every yard (Parkinson, 1973:119). The British Expeditionary Force's retreat to Dunkirk was similarly impeded by fleeing Belgians, carrying everything they could on cars, carts and bicycles. One old soldier remembers trying to pass a farm cart drawn by a single horse; the cart was carrying eight or nine people, all manner of bundles and suitcases, and a grand piano (pers.com.).

This anecdotal evidence serves to remind us that theory and real life do not always coincide.

A study of abandonment at Zuni farming villages has noted that typically, personal items and artefacts associated with transportation were missing from recovered assemblages (Rothschild et al, 1993). Their study resulted in the ability to interpret certain types of deposit and their relationship to abandonment and use practice.

> *"It appears that the study of artifact distribution both inside and outside structures can inform us as to where on the "use-abandonment" continuum the structure lies. The significant variables include distribution (random v. clustered), diversity, density, and size of materials. Density of construction debris seems related to the use state of the structure. As the structure deteriorates, hardware and parts accumulate inside. Stockpiling of building materials produces a different spatial pattern, with piles of like materials (building stone, planks, etc.) concentrated rather than scattered outside the structure....The likelihood of redisposition of small artefacts from various kinds of disturbance makes them rather unreliable indicators of use. The surface scatter of somewhat larger objects, in contrast to their collection in refuse piles, is an indicator of the lack of maintenance of an area" (ibid: 136).*

Tomka (1993) points out, however, that change to different agricultural regimes, or movement to regions where regular crossing of territory is less easy and visits to unoccupied sites harder to achieve, may result in complete emptying of the original site, since reoccupation is not anticipated and this can mimic catastrophe abandonment of the type associated with warfare.

Where abandonment is gradual, the process of disposing of rubbish and those items not deemed to be worth packing can have a certain pattern. Things tend to be dumped in yards, and later become transported to dumps and fields, while items that *may* have a future use are stored under beds, in interior corners, or along walls or fences. These are *provisional* discards.

> *"De facto refuse disposal involves the transfer of materials retaining some functional or other value to the archaeological record during site abandonment without being discarded (e.g. leaving functional items at an abandonment site because of inability to transport the object, haste of site abandonment, or caching activities...)..."*
>
> (Joyce & Johannessen, 1993:138).

These authors note that people planning to move may relax their normal standards and start to dump rubbish in places that would normally be kept clean. It is important, though, to avoid modern assumptions about what items constitute 'refuse' - even in modern society, broken items may be retained for sentimental reasons, to which we could add ritual and ancestral meanings for earlier societies.

In a study carried out by Stevenson in 1982 in the Yukon,

> *"In sites where abandonment was planned and gradual he found little de facto refuse low associations between this type of material and activity areas, and few normally curated items. However, at sites which were rapidly abandoned, de facto refuse, including some items normally curated, was abundant and highly correlated with intended areas of use. When abandonment was planned with anticipated return to the site, considerable quantities of de facto refuse was found. When return was anticipated, de facto refuse was cached to protect it for reuse upon return to the site.*

Type of desertion	**Structures**	**Contents**	**Resulting changes**
Short-distance move/ return anticipated	Structures left intact	Furnishings left in place or removed for use elsewhere	May attract scavengers
Long-distance move/ return anticipated	Steps will be taken to protect houses – sealing of entrances etc.	Caching of valuable furniture/ equipment may result in rich assemblage recovery	May attract scavengers
Short-distance move/ long-term absence or no anticipated return – if house only abandoned	Salvage of structural elements; may be converted to other use	House contents removed; change of tool/equipment inventories	Whole settlement abandoned; insect/ weed infestation
Long-distance move/ long-term absence or no anticipated return		Salvage of some portable equipment/abandonment of heavier/bulkier artefacts	If roofs left, scavenging may take place; if roofs destroyed rich floor assemblages

Table 2: Abandonment – forms and associated actions and results (after Schlanger, S.H. & Wilhusen, R.A.I., 1993

This caching behaviour often resulted in the clustering of de facto refuse relative to rapid abandonment conditions as well as a decline in the direct association of de facto refuse with activity loci"

(Joyce & Johannessen, 1993:139)

and they conclude that

"Stevenson (1982) has shown that material patterning at sites with evidence for rapid, unplanned abandonment should closely reflect the activities performed during site use. However, the research at La Concha suggests that gradual abandonment, especially when reoccupation is a probability, has a much more variable effect on material patterning at domestic sites.....The La Concha research suggests that broad areal sampling at archaeological sites is crucial to control for abandonment processes. Sampling of multiple structures is important since...site abandonment can result in differential treatment of buildings..."

(ibid: 151).

These studies may offer useful considerations for the interpretation of abandonment strategies at prehistoric sites.

The degree of recovery will to some extent depend on in what condition a structure is left; once the roof is removed or burned floor layers are effectively sealed. Where the structure is left to decay with the roof intact there will be eventual deterioration and natural filling. If abandonment is planned, it may be accompanied by deliberate demolition of structures for reuse of elements such as large timbers, fittings, carved or decorative pieces, roofing materials and the like. If posts are removed, the postholes may be filled with non-organic sediments, and their profiles may be disturbed due to rocking out of the timber. Most structural elements will appear in disarray and there will be a general lack of structural integrity. Where a structure has been abandoned without plundering for reuse or in a hurry, the postholes may rot *in situ*, or be burned. Roofing materials may have collapsed inwards, sealing floor layers. In an unplanned abandonment, larger artefacts which normally might have been curated because of the energy investment in their creation may be found where they were left; in a planned abandonment, such items may also be broken up for recycling or removed completely. Brooks (1993: 186-6) suggests that in an unplanned desertion of a structure, there is likely to be a high proportion of refit sequences present, with more artefact classes and more complete artefact sequences represented, and their spatial patterning would be more uniform. Where desertion is planned, there would be less refit sequences and more items are likely to be restricted to the periphery of the floor space or communal areas.

Abandonment as a result of preparation for war is likely to have been *planned*; outlying, vulnerable settlement units would be deserted during a process of agglomeration towards a defensible centre or mutually supporting group of settlements. Clearly, immense problems exist in trying to establish the reasons behind abandonment - the only occasions where preparation for war could be maintained as a cause would be if nearby defensible, agglomerated or defensive settlements could be proved to have been founded or expanded at exactly the same time as outliers were deserted, by people with exactly the same material, ritual and social cultures. Identification of which settlements might have been peripheral to the group needs to take place - criteria such as marginality of cultivable land, distance from water sources, trade routes and denser clusters of habitation, and closeness to possible territorial boundaries (such as rivers, ridges, or entry points into a region) would need to be applied.

xi) Appearance of defensible settlement strategies

"By more defensible is meant the clustering of large numbers of people close together, the ensuring of a good view from at least one site, and line-of-sight communication with one's neighbours...It is likely that these communities were built to be as defensible as was thought adequate for the time, but ten to twenty years later this level of defensibility was inadequate"

(LeBlanc, 2001:28).

LeBlanc's study of settlement in west central New Mexico has identified a major settlement shift occurring between 1264-77AD; new settlements appeared to be more defensive in form than those they replaced, forming large blocks of rooms with high outer walls, carefully sited to secure water supplies, and intervisible. There may be some evidence for a co-operative building strategy. Similar changes occurred in the Grasshopper region of East Central Arizona during the late 1200s. In the Pueblo II and III phases the pattern was of small scattered settlements with a mixed subsistence base that included some hunting. Populations may have been fairly mobile. This pattern lasted into the mid to late 1200s. Despite a rather later pattern of change, the cultures of the region were part of a ceramic network that was similar over a wide area. In the late 1200s comes the first significant sign of aggregation of settlements with the construction of small pueblos, each adjacent to a spring or stream and arable soil. These were focal villages in discrete isolated settlement clusters, each with a 1-2km diameter area. Their construction is associated with a rise in regional population., increased sedentism, and movement of new peoples into the area. Here, the Pueblo IV phase saw the emergence of 10 large pueblos. Their siting was not consistent with the local agricultural potential, nor the best watered lands. Settlements used cliff edges, and views over watercourses, more consistently than access to arable lands. Whilst this region does not produce any strong confirmatory evidence for warfare (just one body with an embedded projectile point) during the period, it is situated in the area known as Apacheria, and was occupied during the period in which the notoriously fierce nations of the Apache were thought to be colonising the area (Tuggle & Reid, 2001). Other defensive strategies were adopted here at the same time (noted below). Defensible strategies for settlements can be subdivided in several further categories - location, morphology, provision of guard posts, creation of buffer zones and intervisibility.

xii) Changes in settlement location

Studies in the southwestern United States have revealed patterns of settlement relocation and abandonment that may be directly attributable to increases in hostilities.

> *"The twelfth-century increase in the intensity of warfare across the Southwest apparently curtailed the use of agriculturally marginal terrain by small, single-family settlements, bringing to an end several centuries of highly flexible subsistence and settlement practices. The thirteenth- and fourteenth-century development of large aggregated settlements, the abandonment of extensive areas, and the clustering of settlements into subregions resembling polities were also very likely outcomes of the increase in warfare"*
>
> (LeBlanc & Rice, 2001:7-8).

Peripheral settlements on the edges of the Sonoran desert were increasingly difficult to defend, for example. As well as settlement aggregation, buffer zones appeared in certain areas, such as that which emerged in the nineteenth century in the Salt and Gila Basins of Central Arizona. From a combined unbroken 60 km. line of Pima and Maricopa settlements in the early part of the century, the mid century picture was of a distribution only some 25 km. in length, with unoccupied regions west of Pima Butte and north of the Salt River, contraction due to the increase in raiding by the common enemies of the Maricopa and Pima.

Two opposing forms of settlement strategy possibly designed to cope with aggression have been observed in the studies of change in Puebloan and contemporary cultures - *aggregation* and *massing*. "In the south the Hohokam used organizational principles that enabled them to *mass* forces from within territorial blocs while retaining a relatively dispersed settlement system. In the central mountains and on the Colorado Plateau populations *aggregated* into very large settlements as a means of ensuring the presence of a sizable fighting force" (LeBlanc & Rice, 2001:8).

The advantages and disadvantages of each form of strategy can be summarised in Table 3 below.

These strategies do not have to appear overnight; people do not readily move away from the places they and their ancestors have farmed for generations. The threats of danger or the difficulties in maintaining living standards have to accumulate over time, and become so overwhelming that the inertia of populations can be finally overcome. The

MASSING

Dispersed settlement pattern
Small settlement units
Absence of fortified centres
Adaptive subsistence strategy
Maximisation of environmental resources
Smaller groups receive share of resources
Smaller groups protected
Stores and stock dispersed
Peripheral units act as sentinels
Peripheral units vulnerable
Some units endangered
Aggressive defence strategy
Alliance structure strategically important
Range of tactical response large

AGGREGATION

Aggregated settlement pattern
Large settlement units
Fortified centres or refuges
Maladaptive subsistence strategy
Destruction of environmental resources

Other groups excluded
Stores and stock concentrated
Possible lookout points
Few peripheral units
Danger is to whole community
Reactive defence strategy
Alliances non-strategic
Range of tactical response constricted

Table 3: Massing and Aggregration Defence Strategies

Stage 1 (*Short duration*)	Initial shift towards more defensive configurations No abandonments No unoccupied zones
Stage 2:	Some site relocation and site reconfiguration
Stage 3:	Change in community size, location, layout Abandonment of outliers Unoccupied zones open up

Table 4: Staged Response Pattern

process may be simplistically modelled as follows *(Table 4)*.

> *"In almost every area where the increase [in warfare] has been seriously investigated, we see a staged response. There is an initial shift in settlement pattern to somewhat more defensive arrangements. This interval is often short, and it is not accompanied by abandonments of the formation of unoccupied zones. That is, settlement locations, configurations, and distributions initially do not seem to be sensitive to defensive needs; they are followed by some relocation of sites, or reconfiguration of sites, that increase defensibility but not by very much. This interval is, in turn, followed by much more dramatic changes in community size, layout, location, and the opening up of empty zones."*
>
> (LeBlanc & Rice, 2001:11-12).

The 'Staged Response Pattern' is evident in studies in the El Morro region of New Mexico and is demonstrated by the following sequence of change (*Table 5*).

During the Scribe S Phase in the mid 1200s, seven communities shared the valley, spaced between 30 and 900m apart, and the size of the villages varied from a handful to up to 40 families. These settlements were all constructed between 1264-1277 during an initial shift to a more defensible organisation. There are good line-of-sight connections between the sites, with one room block being placed at a higher level to facilitate this where necessary.

Twenty years later the Scribe S site itself was apparently attacked - the evidence consists of multiple contemporary burning episodes, and the fact that some rooms were emptied of artefacts, either salvaged or pillaged. The community response was the subsequent building of the Pueblo de los Muertos, Atsinna, Cienega and Mirabal sites, each displaying defensive characteristics including large blocks with high outer walls, which were carefully sited to secure water supplies, and were intervisible. At the same time, other sites were abandoned. Similar patterns and sites are found in neighbouring valleys and areas, with perhaps 20 miles between clusters; there were a few outlier sites, but these seem to have been shortlived. It is noted that these changes were generally maladaptive, with overburdening of local resources leading to nutritional stress, and ultimately population decline and/or dispersal (LeBlanc & Rice, 2001).

Small isolated farmsteads
replaced by
Construction of small pueblos near water and arable in small clusters
replaced by
Construction of large pueblos not consistent with best arable alongside small open settlements and cliff dwellings, observation posts and guard villages across access points

Table 5: Staged Response Pattern in the El Morro region

Similar maladaptation was seen in a study of the Mesa Verde sites of Colorado. Here, oral tradition bears out archaeological observations that change in settlement location led to subsistence stress. The story told by local tribes is that their ancestors were farmers and craftsmen who lived in the fertile valley bottoms; they were visited by savage strangers from the North who they welcomed, but the visits became more frequent and troublesome. The strangers began to raid, steal and attack farms, so the farmers moved up to the cliffs, but could not produce enough food. When they began to starve, they were forced to move away until they found a rocky area where they built fortifications, towers, and reservoirs. But the enemy followed and fought them; the families were evacuated and after terrible battles, the warriors followed them to a new followed and fought them; the families were evacuated and after terrible battles, the warriors followed them to a new place, where they built new towns on unapproachable bluffs (Lightfoot & Kuckelman, 2001:33-6). Castle Rock Pueblo was begun in 1256 and was abandoned by the early to mid 1280s. It is a defensibly sited village around a large butte, one of several such locations that were increasingly occupied by between 15 and 100 families in preference to previous occupation patterns of one or two households on good arable soils. Abandonment of good arable is matched by an increasing trend to secure water supplies and to build enclosing walls and look-out towers. At the end of the occupation aggression is demonstrated by the presence of unburied bodies, some with trauma, and generally increased levels of postmortem neglect.

The use of an aggregation strategy means that everyone has to be involved, because supplies are centralised, and all are committed to defence in a single locale. On the other hand, the massing strategy supports a dispersed settlement pattern which allows maximum environmental exploitation; attackers have more targets, but they are smaller and offer less booty; each has to be attacked separately, slowing the advance. Mobilised defenders have time to assemble (from up to 40km in the Pima case) and can choose their ground for a counterattack. They might lose a few houses, but not their whole stock.

To summarise the above studies, the salient points are as shown in Table 6.

Extrapolation can lead to interpretations about the social and trade relationships of societies adopting these strategies. Aggregation can be seen as maladaptive, creating internal stresses and competition; considerations may include internal competition for resources and growth of dominant settlements and elite groups. It is also possible that aggregation is most likely at the most vulnerable end of the group, in order to enforce fair distribution of resources

Aggregation patterns:

Relatively narrow floodplain, springs:
Small settlement units on best arable soils and water sources
Expansion of production and population
Shift of settlements to defensible higher terraces and use of masonry
Change in arable practice to utilisation of upland dry fields
Abandonment of lower settlements
Development of larger aggregated settlements
Development of fortified settlements at strategic locations

Plateau settlements:
Isolated and dispersed settlement
Introduction of ritual centres
Internal competition leads to some sites becoming dominant and others decline or are abandoned
Peripheral settlement areas and isolated settlements most frequently attacked
Fortified refuges created, settlements acquire defensive postures, peripheries depopulate
Creation of larger ritual/refuge settlement forms with defensive characteristics; also serve as elite residences

Desert region settlements:
Settlements on valley floors
Settlements shifted to high butte sides and crests; valley settlements abandoned
Fortified refuges at strategic places

Massing pattern:
Small, defended settlements dispersed among fields over large area, utilising best soils and water.
Alliance of tribes
Smaller tribe benefits by being allowed to share good land, and from protection
Larger tribe benefits because attacks fall first on the smaller tribe, providing early warning and buffer
Both tribes combine when attacked
Defence based on runner system and readiness to respond
Defenders mass and counterattack
A few farms may be lost but attack is dispersed and slowed
Defenders able to choose appropriate ground and time for counterattack

Table 6: Summary of Aggregation and Massing defence patterns

for those most responsible for defence. The most vulnerable settlement may also become the most dominant, as it may be here that chiefs and warriors congregated.

Massing strategies, however, also have their checks and balances. External groups are reliant on central groups for supply and trade, and have a more precarious subsistence; if there is demographic rise, pressure grows from external groups but when central groups also become more populous, they tend to have more organisation, communication, cohesion and mutual support. Central groups therefore become militarily stronger, and external groups may react by creating strongholds and refuges, although typically these are shortlived. In each case, there must be a trade-off between defensibility and standards of life and growth.

xiii) Changes in settlement morphology

Structures in a settlement under threat may begin to include defensive works, increased storage facilities, and agglomeration (perhaps because of the abandonment of outlying hamlets or farms). In times of sustained attrition structures may show signs of having been unmaintained, individual dwellings may go out of use, the settlement could show shrinkage, or the erection of jerry-built structures created by smaller numbers or less physically strong or able builders. Some of these features have been noted in the studies in the Southwestern USA.

xiv) Appearance of guard settlements

Tuggle & Reid's study of the Grasshopper region of Arizona (2001) identified a number of small sites which they have interpreted as observation posts, with little evidence of storage or occupation. Some villages, too, were placed on knolls on the edge of the plateau. These, they suggest, were possible guard villages at the boundaries of the territory - relatively small, with intensive occupation, in isolated defensible positions. They may have acted as a sentinel system for interior villages necessarily sited near to the best available arable.

A further possibility is that settlements on the fringes of a territory, particularly where trade routes enter a region, might tend to be the largest, most populous and best defended. Such a settlement might hold a population with a greater number of warriors than others, and might, therefore, be associated with acquisition of elite status and control. Thus there are two possible scenarios - a small, strategically sited outpost with minimal storage and occupation evidence, or a large populous settlement with elite features - both of which function as warning and protective sites for their hinterlands. If guard settlements can be identified in the American cases, can they be sought or identified in other places and periods? A correlate which encompasses the existence of guard settlements, therefore, would lead to a focus of research on the extreme edges of settlement territories.

xv) Settlement locations related to line-of-sight links

Line-of-sight links and intervisibility feature strongly in a number of the American studies, typical of which is the case of El Morro, where

> *"independent findings show several patterns: an increase in settlement clustering, settlements arranged for line-of-sight visibility, settlements becoming more defensive over time, increases in site size over time, increased concern for secure*

domestic water, and smaller sites being more defensively located than larger ones."

(LeBlanc, 2001:48).

Determination of line-of-sight links is more difficult in older landscapes with a longer history of utilisation and change, because of the need to take into account factors such as erosion, land use, plant cover, and accessible viewpoints. Line-of-sight links might be of particular importance in a situation where a defensive strategy of massing is adopted, although mutual defence strategies of aggregated settlements would also benefit from such warnings.

xvi) Change in subsistence patterns

We may take as a model a small subsistence farming unit in which some 30-50 people as a maximum might live. It may have an infield and an outfield system, comprising arable, pasture, and meadow, with associated woodland. There will be physical features including water sources, field boundaries, trackways into the field system and trackways to neighbouring settlements, dwellings, storage places, work areas, and workshops.

There may be a sequence of events associated with the preparation for war over a period of time in such a unit, each of which may leave behind some evidence. Outfields may be abandoned early in the period because work parties have become at risk from attack, stock has become more liable to rustling, and/or the community has over a period of time under attrition suffered a decline in its health and size due to restriction of exploitable territory and loss of prime members to violence. It may be very difficult to differentiate between these effects and those associated with crop rotation or seasonal fallow with temporary abandonment of areas of land, and long term shrinkage of the population due to economic, social or health causes.

Where infields are abandoned it may be because of change in subsistence caused by soil depletion, climatic change, community degradation, or external influence, rather than conflict. Evidence for abandonment would include the appearance of typical weeds - clover, cleavers, nettles, burdock, brambles, ground elder, mallow, etc. A brief list of some possible indicator species that may provide pollen evidence to demonstrate land use change can be found in Appendix A. Study of macrofossils, soil chemistry, spores, and pollen may help to elucidate the sequence and speed of abandonment. There may be an accumulations of rubbish in previously cultivated areas, and silting or slumping of field boundaries.

If the people who lived and worked in the settlement leave it, effects should be seen in the surrounding natural landscape. These will include *effects of abandonment*: a possible increase in biodiversity locally, the outgrowth of hedges, the failure of drainage systems and so on. In a long drawn out period of low level aggression, field defences may be developed in the form of increased size and complexity and there may be stronger definition of boundaries, especially between neighbours or on vulnerable sections of the territory.

Abandonment and change in fields and gardens, associated boundary structures and routeways are even more difficult to date, define and attribute to cause. Despite repeated attacks and raids, farmers tend to hang on, year after year; they rebuild, replant and pray that this year, it will be their neighbours who will be hit, not them. It takes a great deal to persuade farming people to give up their land permanently.

The records of the Near East and Greece are full of mentions of military campaigns which often had devastation of the countryside as a main aim. In the sixteenth century BC, a stele of victory at Karnak in Egypt recorded the boast of the Pharaoh Kamose at the conquest of Hyksos - "I shall drink the wine of your vineyard...I shall destroy your dwelling place and cut down your trees" (Pritchard, 1969:554).

In a letter to Assur, father of the gods, around 714BC, Sargon, king of Assyria records that:

"The trunks of all those trees which I had cut down I gathered together, heaped them in a pile and burned them with fire. Their abundant crops, which in garden and marsh were immeasurable, I tore up by the root and did not leave an ear by which to remember the destruction..."

(Luckenbill, 1926: 87-88).

Greek accounts suggest a seasonality to raids and warfare; wheat and barley ripen between the middle of May and the end of June in Greece, and at this time raiders could burn the crops in the fields or reap it for themselves. By late June or July, the crops were gathered in to the threshing floors and barns, easy to seize or destroy. Therefore, "...the Athenian sea-borne forces which attacked the Peloponnesian seaboard usually set sail from Attica at or near the end of June, right after the cutting of the grain. In these raids, small, isolated towns, without good walls, along the coast of the Peloponnese, the Argolid, and northeastern Greece ... were usually targeted...destroying newly harvested produce which was collected inside or near the walls" (Hanson, 1983:34). Devastation of cropland was a well-known and practised strategy of war in these city-state societies.

It was a matter of routine to plunder and destroy farmhouses or hamlets. The destruction of buildings served to disrupt agriculture by denying farmers the use of barns, granaries, byres and mills, and also leave many people homeless, suffering psychological distress and loss of morale. Literature sources suggest that evacuation was commonplace. People removed their livestock, slaves, and where possible food and crops, farm implements and household furnishings, especially food. There are records of military commanders protecting harvesters, providing garrisons and lookouts, and of cavalry playing a major role to protect the harvest. Thucydides and Aristophanes both mention farm implements brought into the cities. Furnishings and metal goods are often mentioned as booty. Sometimes, farmers made efforts to protect themselves - there is mention of traps dug to break horses' legs, flooding of pastures, and the building of barricades.

The arrangement of the countryside itself helped to slow attackers and protect farms. In Attica, many landholdings were quire small and surrounded by stone walls. Many farms were sited on hilly terrain. Efforts to destroy orchards were rarely successful - olive trees are difficult to burn or bring down, and vines tend to recover from quite drastic treatment. Only a really determined aggressor could spare the manpower, time and effort to completely devastate a countryside, all the time exposing his troops to counterattack. Another reason for the ability of Greek farmers to survive attacks was their mode of farming - subsistence agriculture in which productivity and efficiency rested not on capital investment but on intensification of labour by the family and their slaves. There was little of intrinsic value to destroy –

> *" Mud-brick, clay, rubble, and cut stone used in the building of houses, fencing, utensils, ovens, vats, and storage facilities were cheap and plentiful throughout the Greek countryside. To replace such capital losses ... all that was needed was labour - labour that was relatively plentiful, and usually evacuated and kept safe from attack"*

(Hanson, 1983:146-7).

More rarely, defensive measures were taken on a military scales in the countryside. Forts and garrisons in the mountains delayed and sometimes halted the enemy before they could reach the farmlands. A famous attempt at defence occurred in 378BC when Agesilaos invaded Boeotia. He proceeded to ravage the land as usual but, according to Xenophon, "he found, however, that the plain and most valuable sections of their land had been enclosed with both a trench and surrounding palisade" (quoted in Hanson, 1983:69). The Thebans within the defences matched Agesilaos's attempts to find a place to break through. "They were aided by cavalry sorties which issued out of exits in the walls and harassed the surprised Spartans. Agesilaos, however, attacked at sunrise one morning, burst through the unguarded sections of the walls, and ravaged the plain" (ibid:70).

Some time in the 4th century BC a long field wall (c4360m) was built in the interior of Attica across a gap in the mountains of Igaleos and Parnes, called the Dema Wall, intended to cut the direct access to the Athenian plain and croplands, and the 11 km long Antiforitis Wall was built in 313BC to separate the interior of Boeotia from the small coastal plain of Salganeous, to guard croplands, (Diodorus 19.77.4). Hanson concludes that for often conservative agricultural societies, the main damage done by warfare is psychological. "Devastation is frequent and universally feared, yet serious long-term economic damage does not automatically result" (1983:150). Nevertheless, it has often been used as a military tactic, from the deliberate devastations designed to reduce national faith in the government of France perpetrated by the English armies of the Middle Ages to British burning of grain in the fields during the Boer War (Pakenham, 1979) and the devastation of the Shenandoah Valley in the summer of 1864 when General Sheridan set fire to the fodder and grain fields.

One of the problems in identifying this type of event is that abandonment and shifting of agricultural use may be employed as an effective subsistence strategy. Shifting settlements allow maximum long-term utilisation and production levels - it is not unusual for communities to rotate the village as well as the crops. There is a need to look at the regional picture as well as individual sites. A study of the Mimbres region in New Mexico revealed that abandonment of one type of settlement was matched by the appearance of another form, and continuity as well as variability in artefact and architectural styles. "These findings point to the importance of considering village-level depopulation as a process, and examining when and why it sometimes leads to regional reorganisation and at other times to long-distance movement" (Nelson & Hegmon, 2001:230). Like other writers, they call for large scale research to study broad patterns of migration and settlement shift.

Subsistence change as a correlate of warfare is particularly fraught with problems of identification and is unlikely to be useful in isolation; however, when combined with the presence of a high proportion of confirmatory correlates, it may act to substantiate such an interpretation.

xvii) Change in trading patterns

Changes to routes of communication are also likely; some routes will go out of use. Trackways into field systems may show evidence of silting, broken surfaces, or blockages, or of a change of use such as cattle or sheep movement, resulting in deeper cutting and surface changes. Trackways to neighbouring settlements may become abandoned, or new tracks may be created, demonstrating possible change in exchange or political focus.

Trade relationships with allies are likely to be strengthened in the preparatory period, whilst relationships with potential enemies or with trading partners divided geographically by intermediate potentially hostile groups or by vulnerable routes will probably diminish. The material culture of a community under threat may display an increase in weapon stockpiling, and a decrease of exotic goods as trade becomes more difficult. Tracing of the origins and numbers of imported and exported goods from a region may help to define patterns of alliances and interruptions to trade relationships; the general lack of close dating, however, may make it difficult to ascertain whether changes in pattern are due to realignments preparatory to conflict, or resulting from it.

xviii) Unoccupied zones developing

Caesar's report of the tactics of the Suebi has already been mentioned - maintaining a no-man's-land around their territory served to underline their image as fierce warriors, contributing to both internal pride and identity and external prestige. In less belligerent societies, the opening up of unoccupied zones may be a form of fence-mending between communities desirous of avoiding conflict, or this may reflect a process of attrition aimed at outlying settlements and activities that has resulted in contraction of utilisation areas. ice and LeBlanc (2001) note the repetition of the occurrence of unoccupied zones in several of the studies they have considered, many of which are attributed to a process of attrition. However, the effects of tenurial patterns (as in the Swedish example above) or change in subsistence strategies may create similar evidence. Once again, this correlate is only useful in the consideration of the occurrence of war when it can be seen to accompany a sufficient proportion of others.

Functional Correlates

xix) Burned sites

Burning is by no means the only way in which settlements and lifestyles can be destroyed by an enemy, but it does tend to be one of the more archaeologically visible methods, and is a very common method of attack, particularly in regions where the dominant architecture includes timber and thatch. In such areas, of course, the incidence of accidental fires is also likely to be high, and intentional destruction of structures by their erstwhile occupants (for reasons of ritual or hygiene) is also frequently accomplished by burning. The problem, therefore, is to be able to differentiate between these different types of fire.

This question was addressed by Glen Rice in his study of warfare in the Salt and Gila River systems of Arizona. The village of La Ciudad around 1000AD displayed a dispersed morphology, with wattle and daub houses, in groups of between two and five arranged around a yard, each cluster of yards related to a cemetery. The houses were strung along edges of four canals with groups of clusters (loci) over an area of approximately 1.25km sq. Distances of up to 500 m occurred between the loci which were used as arable. There was evidence for periodic attack over a 300 year period in the form of burning episodes at the edges of the settlement; houses in the centre of the area were rarely burned. Those on the southwestern edge suffered 38-63% burnings. The earliest houses at La Ciudad c 750AD were scattered and relatively isolated. Between 750-850 there were 12 scattered houses, of which eight had been burned. Between 850-950 more houses were constructed and the courtyard pattern emerged. A ballcourt was built; burnings became much more rare, with just 33% of 17 houses at the Belleview ballcourt locus burned in the whole period 940-960. However, soon after about 960, and again around 1030, over 50% of the houses in the settlement were destroyed by fire. Solitary outlying houses in the surrounding countryside were burned at the same time. The less defensible northern section of the town began to contract, and was wholly abandoned by 1100. If, as has been suggested, burned houses are the result of accident or forms of mortuary ritual, it would be expected that there should be a fairly even distribution of such buildings over time and space. At La Ciudad, that is clearly not the case - burnings are largely restricted to limited periods of time, and are concentrated in the outer, least defended, parts of the settlement, and "large assemblages of ceramic vessels, implements, household items, and metates and manos [maize grinding equipment comprising a heavy stone dished quern and large rounded rubbing stone] on the floors of the structures suggest the burning was a precipitous action with expensive consequences" (Rice, 2001:324).

The peak period for burnings (1000-1150) coincides with the first desert fortified refuges. Post-1300 the burnings decline, at the same time as walled compounds appear in the canal region and depopulation starts in the desert (Rice, 2001). In other locations, with more scattered occupation, it would be much harder to be able to define the difference between accidental and deliberate burnings, although burning for ritual purposes may still be identifiable through a consideration of the pattern of artefact abandonment within the remains of the structure.

xx) Unburied bodies/weapons trauma (human and animal)

This topic has been discussed at various points earlier, both in relation to the validity of skeletal evidence as confirmation of conflict, and in relation to the effects of warfare upon domestic and wild animal populations. It is likely to be extremely rarely that data of this sort will be found archaeologically in secure contexts. Where it does occur, alternative explanations of accident, disease or murder need to be assessed with care.

xxi) Weapons damage

The recognition of use-wear damage on weapons, and the alternative explanations for such damage, have been discussed in Chapter Two. Ritual damage may be identifiable in some cases (York, 2002); damage inflicted during training would be less easy to differentiate from use-damage in actual war. Nevertheless, whether damaged in training or in battle, such damage at least indicates a level of preparedness for conflict, and the presence of functional warriors.

Resultative Correlates

xxii) Change in subsistence patterns

Subsistence pattern change as a result of war can occur in several different ways - through dislocation of populations, through changes in foci of production linked to market change, dominant group practice and/or ideology, and through the exigencies that particular forms of destruction place upon recovery.

The effect of the dislocation of populations in the former Yugoslavia has been mentioned above: abandonment of small subsistence farms and reluctance of younger people to return to isolated and poorer areas. The Roman state found itself having to deal with the latter problem in its earlier phases, as time-served soldiers chose to retire to urban locations rather than return to farms, a problem at least partially solved by the setting up of provincial coloniae and the granting of land to veterans in conquered regions.

Roman practice also gives the example of the creation of imperial estates in conquered territories, and the imposition of new field systems, particularly centuriation, examples of which can be found in Gaul and Spain.

Relative concepts of wealth and worth may result in change; a victor whose cultural tradition vests status in the ownership of stock may impose pastoralism on a previously arable landscape, for example. Market change, whether in location or demand, may also significantly affect subsistence patterns, with changes in production reflecting changes in demand from exterior trading partners.

Subsistence practices may be forced to change in the short term because of the immediate effects of hostile action - killing or driving off of breeding stock, destruction of supplies of seed corn, destruction of water sources or drainage systems, destruction of trees, destructions of processing or storage facilities, destruction of means of transport etc. In some cases, alternative strategies adopted in the short term to overcome these problems may become embedded and previous strategies may never be resumed.

In post World War II Britain, while many plots created on marginal land as part of the 'Dig For Victory' campaign were quickly abandoned after peace was achieved, the hedgerows between small traditional fields which had been removed were, in most cases, never replaced. The increased mechanisation of rural production introduced during and immediately after the War favoured these larger fields, and the process of removing hedgerows not only continued, but accelerated. Prairie-style fields in East Anglia may trace their origins in part to Roman drainage schemes; the Romans required corn to feed the legions and new urban populations. They changed the landscape to achieve this, possibly creating imperial estates in some places.

In prehistoric contexts, subsistence change may not have been particularly drastic or pervasive, due to more limited scales of conquest and, perhaps, to a general similarity of subsistence practice. However, there are some indications of large scale subsistence change in prehistory, and the effects of war must at least be considered when attempts are made to explain them.

xxiii) Change in trading patterns

After hostilities, trade is likely to be affected geographically, quantitatively and qualitatively. New trading centres and routes may be imposed, focused on the home territory or other conquered lands of the victors; older trading partners may become inaccessible as distribution routes are interrupted or they too are vanquished.

The quantity of imports to a region may be affected by competition from new sources accessible to incomers, by supply and demand from new regions brought into the trade network, by the ability of conquered populations to pay for imports, and by their willingness to adopt new or different styles of artefact. Exports will be similarly affected; the productive capacity of a conquered, diminished or dislocated population may be severely curtailed, and their traditional local products may lose value and status. The pattern of dissemination of trade goods may be altered within the society, and the rate at which different sectors of a community gain access to imports may vary considerably, depending on the degree of dislocation of pre-existing internal exchange mechanisms.

Change in artefact style is a cornerstone of archaeological investigation, but the extent to which such change may depend on the results of hostilities requires more study, although some work in this area has been carried out [for example, Barth (1969) and Phillipson (1974)].

xxiv) Trophy collection and memorials

Deposits such as at the Hjortspring and Ribemont-sur-Ancre sites have been discussed earlier, as have the deposits of human skulls at Ofnet. War memorials are present in many forms in most modern societies, and certainly the Romans were energetic builders of such monuments - Trajan's Column, the prows of captured ships displayed in the Forum, and a multitude of triumphal arches bear clear witness to this. What form such memorials might have taken in earlier societies, if they existed, is unknown but worthy of some speculation by future researchers.

At the site of San Juan Mogote, a large farming settlement in Oaxaca, Mexico, whose defensive palisade has been dated by radiocarbon to 3260-3160 BP (210-110 BC), evidence cited for warfare includes the burning of temples and houses, an inscription recording a captive, and a skull rack. (Flannery & Marcus, 2003). Many enigmatic monuments in the European landscape might owe their construction to commemorative impulses, some of which might represent victory or loss in conflict, lacking written or iconographic confirmation, it is now impossible to establish the meanings of these stones, mounds and other features.

xxv) Iconography possibly related to warfare

Under this heading might come not only the forms of communal monuments discussed above, but the records of individual warriors in the form of decorations of weaponry and other artefacts, graffiti, or the appearance of artefacts depicting enslavement or sacrifice of enemies. Such artefacts are known from the Graeco-Roman and Aztec worlds, but are likely to be hard to identify, if they exist at all, in prehistoric European contexts, although the warrior-stelae of the Alpine regions might possibly lend themselves to such an interpretation. *Proof* of such meanings is probably impossible.

xxvi) Deviation in sex ratio

In the aftermath of conflict, the sex ratios of surviving populations may be skewed. There may be a missing generation of young and adult males, killed in combat, assuming most fighters were male. There may be a missing generation of young and adult females, seized as booty. And thirdly, young and adult people of both sexes may be absent if taken and sold elsewhere into slavery. The difficulty faced in assessing this correlate is that rarely do sufficient samples of populations appear in the archaeological record to be able to make close population estimates. A further problem arises if societies commonly practise differing mortuary rites for different sections of its population. Skewed sex ratios have been noted in many different instances of cemetery populations, and these have sometimes been attributed to causes such as gender-preference infanticide or gender-preferential rearing techniques for infants. The possibility of sections of the population missing as a result of war is rarely considered in the literature, but perhaps, if other correlates are identifiable, this omission deserves rectification in some instances.

xxvii) Depopulation

Large scale depopulation as a result of war has occurred frequently in the past. The examples of Gaul after the Roman Conquest and of parts of the former Yugoslavia have been discussed earlier. Depopulation following warfare can occur either as a result of massacres, or enslavement, or hunger and disease, or flight of refugees, or of forcible removal of populations. It can also occur as a result of epidemic disease, natural catastrophe, and consensual population movements. Once again, it is only a combination of context and confirmatory correlates that would enable identification of depopulation as a result of war.

xxviii) Unoccupied zones developing

Protective no-man's-lands have been mentioned already; unoccupied zones as results of war might result from the forms of depopulation listed above.

xxix) Population movement

Caused by the factors listed under the heading of depopulation, populations displaced from their point of origin could possibly be identified through various forms of analysis of trace elements and heavy metals in their bones. This technique has been used successfully for several recent discoveries, such as the bodies discovered near Amesbury in 2004 whose bone chemistry suggests a home in the Prescelli Mountains vicinity in Wales. Larger scale sampling of cemetery populations for such analysis might be a strategy that could be included in research designs where population movement is suspected.

xxx) Abandonment of developed arable

In most instances, farmers do not readily abandon their fields. Too much effort will have been put into clearance of trees and rocks, building field boundaries, improving the soil and creating proper drainage for any hasty decision to leave to be entertained, and even if one farmer decides to give up and move, it is very unlikely that this pattern will be repeated across a region. Abandonment of developed arable depends on very few effective causes - soil exhaustion, natural catastrophe, severe and long-lasting climatic change, major pandemics that decimate the workforce, or large-scale political and social changes that may include warfare. It should be possible to establish, through the use of soil, floral and molluscan analyses, whether soil exhaustion or climatic change are implicated in abandonment.

There are possible problems associated with crop rotation and fallowing, but detailed sampling techniques may be able to differentiate in these cases, which should generally represent short-term abandonment.

Comparison of demographic statistics before and after abandonment should be able to confirm or eliminate change due to pandemics - as, for example, change in agricultural practice in the later 14th century AD. Social and political changes are much harder to establish.

xxxi) Changes in settlement location

Post-conflict, settlements may be relocated for a number of reasons. Control of conquered populations may be a concern for the victors, as has been cited as the cause for the foundations of towns such as Dorchester and Canterbury following the Roman Conquest in Britain. Abandonment of vulnerable settlements and relocation to more defensible sites can occur during the process of war and preparation for future conflict; this has been discussed above under Preparatory Correlates, and continues to be a possibility at different stages of conflict relationships. Depopulation as a result of war may lead to amalgamation of populations in new locations that may be considered more viable; psychological effects of war may lead to the foundation of new settlements free of painful associations, ghosts and the evidence of defeat. Finally, forms of ethnic cleansing can force elements of a population to relocate, possibly to marginal lands, to escape victimisation.

xxxii) Changes in settlement morphology

Change is often represented in the form of new installations in the landscape, and recent work in Northern France has produced an example of militarisation of a captured zone following the Roman Conquest, namely the strategic axis represented by the Scarpe valley in the tribal lands of the Atrebates.

> *"The excavations undertaken at Actiparc offer the possibility of concretely following the measures taken by Rome to ensure mastery of the upper valley of the Scarpe. The construction of a small fort on the one hand and an access road into the Celtic domain on the other mark the beginning of a complete restructuring of the countryside conducted under the aegis of the military authority"*
>
> (Blancquaert , Prilaux et al, 2003:22).

Excavations have demonstrated that the military zone extended to encompass a palisaded supply depot, an enclosed Gaulish settlement and area of production, facilities for the troops including a bathhouse, and a large (3ha.) open space possibly for markets or parades. A north to south road crossing the area was also palisaded, forming a restricted access provided with checkpoints. Cemetery evidence included a magisterial chair and a rich array of exotic goods dating to the Augustan era. "This site where military presence and economic activity were tightly implicated is thus an astonishing testament to the orders given by the Roman authorities to pacify the region, but also to facilitate its economic development" (ibid:23).

Further historical examples of this trend by occupying forces include the building of timber and stone castles and the laying out of market places and burgage plots by Norman lords in the century or so following AD1066. A feature of the Incan conquests in Andean settlements was the construction of large storage facilities for the stockpiling of tribute goods, and the improvement and extension of communication systems to enable that tribute to be delivered to centres such as Cuzco. At San Juan Mogote, Mexico, the defeat of the local population is marked by the appearance of a fort built by the victorious Zapotec invaders (Flannery & Marcus, 2003).

There may be hierarchical changes in dwellings (increased differentiation occasioned by the arrival of 'foreign' control), and destruction of defences to prevent use by rebels.

So the changes that might be expected could include elite military installations for the control of a population, new trading and commercial installations for the control of redistribution, and storage facilities for the collection of booty and tribute. These could be differentiated from pre-existing installations by reference to further correlates such as status artefacts, exotic items, and changes in patterns of redistribution and storage.

xxxiii) Changes in biodiversity due to human relocation

Biodiversity can be strongly affected by anthropogenic factors associated with warfare. In abandoned fields and settlements, biodiversity will increase - the restricted range of animals and plants maintained by human agricultural preference and practices will be replaced by a much wider range of species, which will invade and ultimately alter the local range of habitats. In regions into which displaced or refugee populations move, biodiversity will decrease - initially, perhaps, as a result of depredations caused by desperate searches for food, fuel and the means to build shelters, and later by the effects of new agricultural and settlement foundations. That such changes need not be solely traceable to conflicts goes without saying.

xxxiv) Changes in health profiles/standards of living

People who have suffered defeat or forcible removal, or have been the victims of endemic attritional attack, are likely to suffer from a decrease in living standards, perhaps due to loss of arable land and stock, or the loss of adult workers (dead, enslaved, or removed from the workforce to a protective role as warriors or sentinels), or to the need to pay tribute. This effect has been often mentioned in literary accounts of historical wars, and has been noted archaeologically in studies such as Crow Creek (mentioned in more detail in Chapter Two). Where available, pathological studies of populations may reveal changes in health patterns; environmental sampling at habitation sites could identify reduced standards of hygiene and building maintenance.

xxxv) Artefactual and cultural change

If people have been defeated, the results may include lower standards of production, a smaller range of artefacts, an increase in re-use and mending, fewer new items of the original cultural form, and the appearance of 'foreign' artefacts and styles; there is some evidence for the appearance of 'resistance' styles in locally made artefacts, as a form of identity maintenance under the imposition of a foreign cultural and physical power (Hodder, 1978). This type of evidence is difficult to assign to close periods of time, as Hamilton (1995) has clearly demonstrated, but in some cases it may be the *only* evidence for a regional 'change of management', all other symptoms being too ephemeral to leave an archaeological trace. Change in material culture is a constant in archaeological accounts, and there are many possible explanations for its occurrence - warfare cannot be discounted as one of them.

xxxvi) Changes in ritual and burial practice

A feature of the process of war may be the interruption of traditional ritual and burial practices. The causes of this may reflect deliberate destruction of sacred sites by invaders - a practice that achieves demoralisation of the population - or the rendering of traditional ritual or burial places inaccessible, either through fear of attack or proscription by conquerors. Deliberate destruction of a place of worship should be archaeologically visible but cessation of use could reflect change in belief structures and practices for many other reasons. A feature of burial practice that has been associated with war is the abandonment of remote burial grounds in favour of burials in or close to settlements - the decision may reflect a desire for rituals to be carried out in more safety, or constrained circumstances may cause rites to be carried out with less elaboration and ceremony.

xxxvii)Territoriality developing

An increase in conflict and attacks may create the circumstances where a group of settlements forge alliance and mutual defence links against further assault. This strategy might be successful, ending the danger and preventing escalation of war; it may be partially successful resulting in the formation of bigger groups of combatants and larger scale conflicts, but with violence occurring more rarely. Alliance formation for protective reasons is likely to be accompanied by trading partnerships and focusing of redistribution patterns within membership of the alliance, leaving an artefactual trace. Co-operative defensive works might also be undertaken by the larger workforce made possible by the combination of effort of a number of settlements. These processes might also be accompanied by a social and psychological shift in attitude - the creation of 'them' and 'us'.

The allied Pima and Maricopa tribes were often involved in warfare with raiding Apaches or in battles with neighbouring Yuma and Mohave territories, a pattern of behaviours which lasted into the 19th century. In the 19th century, villages consisted of strings of houses near to clusters of fields, with an average of 40-70 yards between the houses - villages could be strung out over several miles. The village unit equated in Pima society to those sharing a particular canal. By this time at least eight canals were watering 12,670 acres. The Pima and Maricopa people speak different languages and have different cultures, but perforce became allied against Apaches, Mohave, Yuma and Yavapai. Historical sources record a form of war in which raids involving five to ten people occurred every few days, and larger groups once or twice a month. Women and children joined in. Raids were mainly mounted in order to steal, but deaths could result. The raids led to a 10 mile (16 km) contraction of Pima territory in one decade through a process of attrition. War parties could number up to c 200 strong. Pima and Maricopa children trained for battle from the age of nine or so. In 1852 the population consisted of c 4,200 people of which 1,000 were warriors, and 3-400 were constantly deployed against the Apache, mostly as sentinels, while women worked the fields. Sentinels were also used at night. Runners spread news of approaching attackers from village to village; they could extend the defensive network for 40 km. Warriors would run to meet at a common point while the raiders would be tracked, and when sufficient numbers were available, a counter attack would occur. The Maricopa were the first villagers to be attacked, and thus acted as an early warning system for the Pima, who would then come to the defence of the Maricopa. The Maricopa suffered more loss, but were a small tribe, and the relationship with the Pima gave them access to the canals and good lands they could not have held on their own.

The earlier archaeological evidence for the region demonstrates similar strategies. Farming relied on canal systems enabling irrigation agriculture on the desert fringes. People sharing each system fought each other, but also fought together against common threats, using bows and hand weapons. An attempt to gain access to canal headgates seems to have been a common reason for opening hostilities; these headgates were limited in number, so the way was open for competition and conflict. All the people living on a canal would have to be involved in protecting its headgate and territory - defending an ecological niche. They also had to co-operate to manage water distribution. Archaeological confirmation for the picture includes the identification of pottery distribution clusters which relate only to the canal systems, even if other settlements were physically closer.

People living in the surrounding desert areas were not organized into any territorial unity; agricultural shortages amongst them were more frequent and so they would have sought food from canal settlements, through trade or raiding. From AD900-1100 desert settlements grew larger and more numerous, and the scale of demand rose. As desert settlements grew more needy, canal settlements grew more populous and able to field large groups of fighters. By the late 1100s a period of war began; initially the canal settlements more often victorious. Desert populations reacted by building fortified retreats on high hilltops or fortifying compounds, but by the late 12th or 13th centuries had been driven out, and in turn dislocated settlements further away (Rice, 2001).

Territoriality has many potential causal origins; a rise in levels of recognition of local ethnic, familial, linguistic, economic or protective identity, or any combination of these factors, may be accompanied by the rise to power of local controlling elites, and development of specific trading partnerships related to access and communication routes and barriers in a region. Any or all of these factors may also be implicated in the process of war; maintenance of defensive or militarily acquired territories is of prime concern to those caught up in conflict, but the archaeologically visible evidence is likely to be ambiguous without supporting material.

Conclusion

A combination of appropriate 'signifiers' of the types discussed above might help to improve the validity of interpretation of abandonment and change as caused by war; conversely, the presence or absence of some 'signifiers' could help to prove that warfare was not a significant feature in the abandonment of a settlement.

Suitable techniques and data are not always available in all of the locations and forms of the clues discussed above, and evidence recovered may be very limited. However, historically too few excavators have taken their interpretations for site abandonment to the level of a developed programme of research into this aspect of a site, particularly in regard to the destruction, abandonment or change experienced in the wider territory associated with a settlement - its fields, woods, gardens, and communications.

The development of a list of testable correlates relevant to a period or location, selected from those discussed above, may assist interpretations or re-interpretations of the possibility of war being a factor to explain behaviour and change.

It is proposed, in the next section of this paper, to attempt to apply the list of correlates to a number of case studies in order to test the limits of available data, to indicate areas of useful additional research, and to speculate on the ways in which the current state of information could be interpreted if warfare had been a factor in the times and places discussed.

THE CASE STUDIES

Case Studies Introduction

The three case studies presented here were selected to test a range of potential interpretations about the occurrence of warfare. Gallia Belgica represents a regional study of an area in which the first recorded war in Northern Europe took place according to historical records. Many aspects of the Julian conquest are not, or only peripherally appropriate to this study - the involvement of a state level apparatus of administration and finance, the presence of a professional trained army, and the instigation of the war for distant political motives. However, the Gauls themselves were, in fact, prehistoric, and little record of their experience of the conquest survives. The record relates to Roman perspectives and interests, not to those of the conquered population. The historical information also fails to record much of the types of information and evidence that archaeology may be able to recover related to this period of conflict. The Gallia Belgica study is thus included to provide a form of control scenario, one in which warfare provably happened in the historical record, but also to test the application of correlates in a conflict that happened at the very end of 'prehistory'.

The 13th and 14th centuries AD in East-Central Arizona saw a period of climate change, subsistence stress and social change in the Hohokam communities of the region. This study concentrates on a series of settlement groups along an east-west river system which have been the subject of a great deal of research and study. There is a bias in the literature of these studies towards an acceptance that warfare was a significant factor in the observed changes in the region. The people of these communities were subsistence farmers with a low level metal and lithic technology, living in permanent villages, and with extensive trade networks. In this sense, they may be comparable to the communities of the Middle Thames area in the Late Bronze Age.

Nowhere in the literature is there any emphasis on warfare as a major influence in the observed changes of settlements and communities in this region, although the presence of a great deal of weaponry has been recorded. A number of settlement clusters along the east-west river conduit have been investigated, with varying degrees of completeness. These people also lived in villages, practised subsistence agriculture and had a similar level of technology and trade to the Hohokam communities. The Middle Thames in the Late Bronze Age is a region in which warfare may be a *possible* factor in the observable changes recorded through archaeology.

The decision to test the correlates against such geographically separated regions and chronologically divided periods was intended to provide as wide a comparison as possible to assess the universality of the methodology. Further case studies may be appropriate in any future assessment in order to refine or refute the methodology in different locations and for different time periods.

Case Study I - The Middle Thames Region in the Later Bronze Age

The study area stretches 50 km west to east, and is approximately 23 km at its widest north to south; it is bounded by the eastern mouth of the Newbury Gap (at approx. SU5560) to the west, and the West London conurbation at approximately TQ0970 to the east, following the Thames Valley and its meanders (*figure 11)*. The northern and southern boundaries are generally delineated by the rising ground above the lower river terraces following the 50m contour.

Topographically, the area comprises a large relatively shallow valley with an east-flowing river and numerous tributaries entering from north and south. The highest point of the River Thames in the study area is at approximately 49m OD, and the lowest approximately 12m OD. The terraces rise to a series of plateaux and spurs rarely exceeding 90m OD in the west and 70m OD in the east with, for the most part, gently rounded profiles. The higher ground consists of Plateau Gravels or Chalk, whilst the valley bottoms are made up of a series of gravels, clays and brickearths of Reading Beds, Bagshot Beds, London Clay and associated forms, with overburdens of alluvium (NERC/IGS, 1960).

Present land use consists of a number of urban settlements, often at or close to the confluences of the various tributaries, interspersed with zones of light industrial parks, gravel extraction sites, reservoirs, some large settlements and a number of smaller rural or dormitory settlements of a suburban character within areas of agricultural use (mostly pasture and some market gardening).

Heathrow Airport covers a large area of the north easterly section of the study area.

The region has a relatively high population density and continuous occupation and exploitation certainly from the Roman period onwards. It has been affected by some major hydrographic works (such as the formation of the Kennet and Avon Canal and the West London reservoirs) and by motorway and road construction, particularly the M4 and M25. Flood control works and channel dredging have also affected water tables and flood plains, sometimes dramatically. Urban expansion has greatly increased during the last 150 years; previous historical and social factors include turnpikes, forestation inclosures and emparkments, notably at Windsor.

The pace of modern development has meant that much of the area has received a great deal of archaeological attention in recent years, although this is by no means universal or comprehensive. It must therefore be stressed that any conclusions regarding apparent patterning of prehistoric features must remain conditional upon future research.

The picture of the area in the Later Bronze Age has become much clearer in recent years, as a result of the large number of developer funded excavations and surveys that have been carried out, as well as several important archaeological investigations into aspects of the period within the region. Attention was drawn to the Middle and Lower Thames area in the Later Bronze Age in the first instance by the very large number of metal tools, weapons and other artefacts that have come to light in or close to the rivers of the region (York, 2002). A major study of field systems along the Thames has been completed (Yates, 1999, 2001, 2007), and a large-scale research investigation at Runnymede Bridge (Needham, 1980, 1986, 1991, 1996) has been undertaken, in addition to the many small- and large-scale rescue excavations that have taken place.

A pattern has emerged of major landscape change beginning in the Middle Bronze Age; the centre of power for Southern England from the Neolithic period had been Wessex and the Upper Thames valley, but by 1000BC, the settlements of the Middle and Lower Thames were becoming richer and more important. Yates (2007:3) believes that the Middle and Lower Thames were well placed to benefit

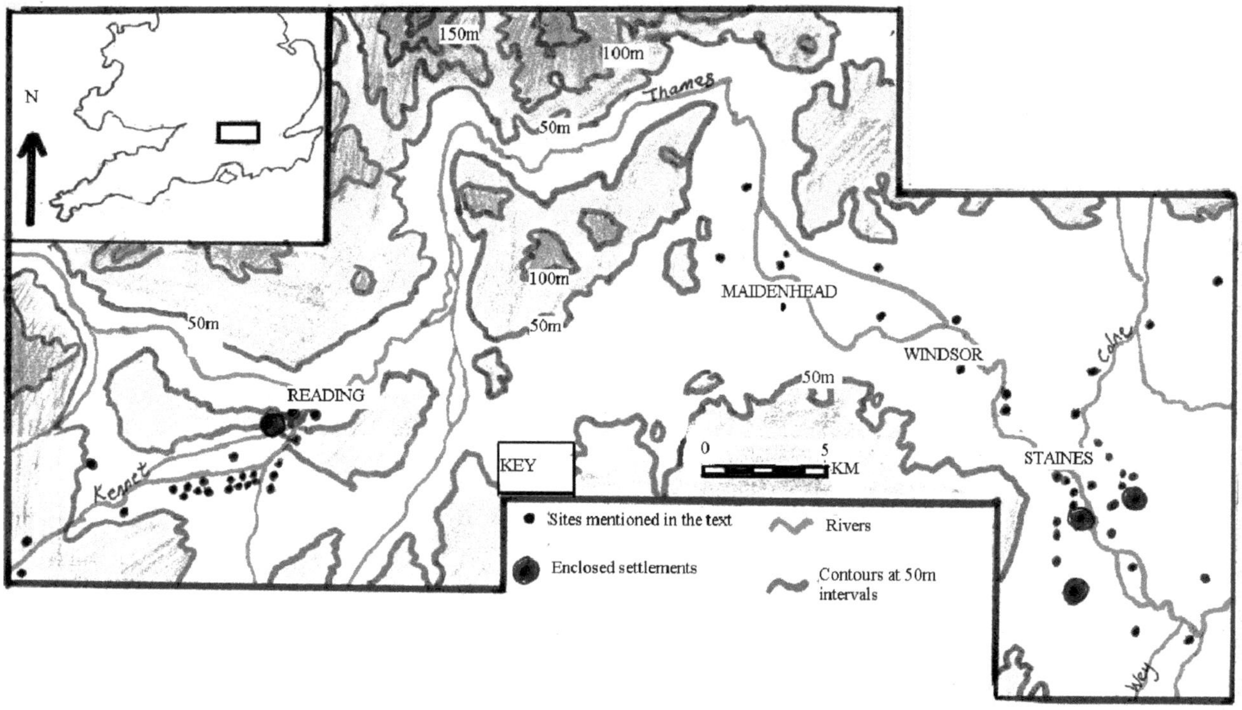

Figure 11: Sites in the Case Study Area

both from agricultural intensification of the relatively narrow farmlands on the river terraces, and, perhaps more importantly, from long distance exchange networks. Northover (1982) has noted, based on a study of impurities and alloys, a 'dramatic' shift in metal distributions from the older sources in Ireland and Wales to a stronger reliance on continental European material in the Late Bronze Age. The Thames Valley has a much greater number of metal goods dating from the Later Bronze Age than anywhere else in southern England The boundary for ceramic distribution between Wessex and the Thames Valley appears also to be well-marked, "with a sharp cut-off along the edge of the chalk overlooking the Thames valley and conform[ing] to the line of the Berkshire Grim's Ditch" (Barrett, 1980:82).

The territorial rearrangement of the landscape continues within the region; a number of settlement clusters can be identified along the course of the Thames, most centring around a confluence with one of the major tributaries. These clusters are, from west to east, at Lechlade, Wallingford, Reading, Windsor/Maidenhead, Staines/Heathrow, Carshalton/river Wandle, along the river Lea, Hornchurch, Mucking and along the shores of the Thames estuary (Yates, 2007). The settlement evidence coincides with concentrations of metal finds and with major groups of field systems. Three of these clusters lie within the study area - the Reading group at the confluence of the Thames and Kennet; the Windsor and Maidenhead group, and the Staines/Heathrow group centred around the confluences of the Colne and the Wey, which includes the major site of Runnymede Bridge.

In addition to the sites identified as settlements, or with evidence of occupation, there is a large number of recorded features of other kinds (field systems, ditches, waterholes and artefact find sites) attributed to the Later Bronze Age, and a further list of sites identified as crop marks or other forms of indication for which dating is, as yet, unavailable (Yates, 2007; Longley, 1976; Johnson, 1975).

The Kennet-Reading Sites

(numbers in brackets indicate sites identified on Fig. 13 below)

SU 6067	Aldermaston Wharf **(2)**
SU 6169	Ufton Nervet
SU 6266	Raven Hill
SU 6270	NE of Bostock Lane, Englefield
SU 6669	Green Park
SU 6770	Field Farm, Burghfield
SU 6770	Field Farm II **(3)**
SU 6770	Field Farm 3 **(3)**
SU 6769	Amner's Farm
SU 6769	Wickham's Field (M4 Services)
SU 6769	Field Farm MSA **(3)**
SU 6869	Pingewood Area C
SU 6869	Moore's Farm, Pingewood **(7)**
SU 6869	Knight's Farm Subsite 4 **(5)**
SU 6870	Knight's Farm Subsite 3 **(5)**
SU 6969	Pingewood, Burghfield **(6)**
SU 6971	Anslow's Cottages, Burghfield **(4)**
SU 7069	Smallmead Farm

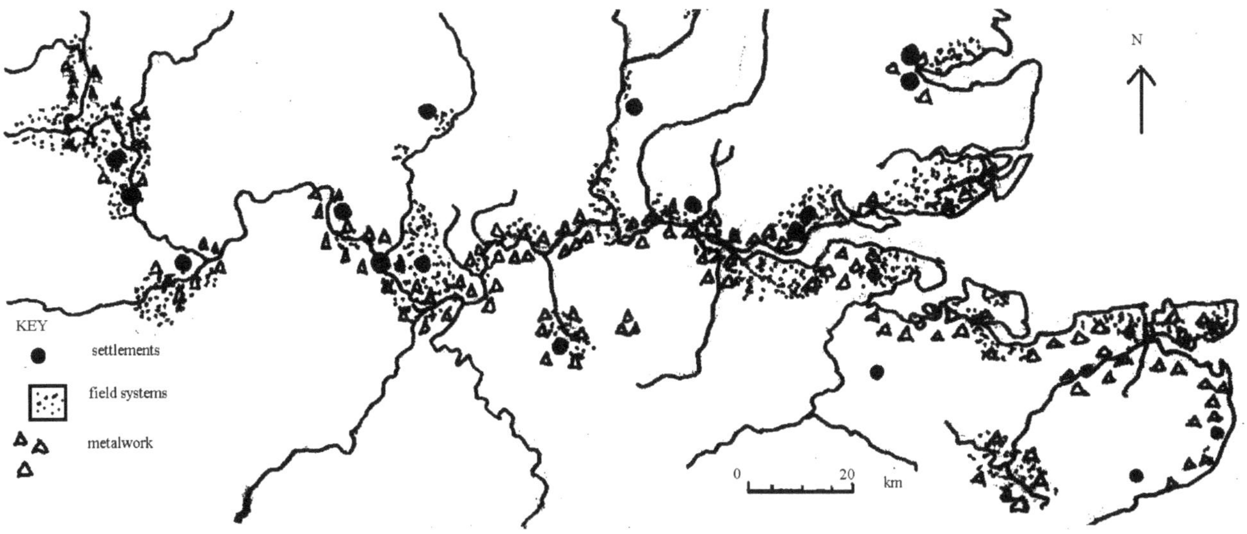

Figure 12: Later Bronze Age metalwork, fields and enclosures along the Thames Valley (after Yates, 2007:112)

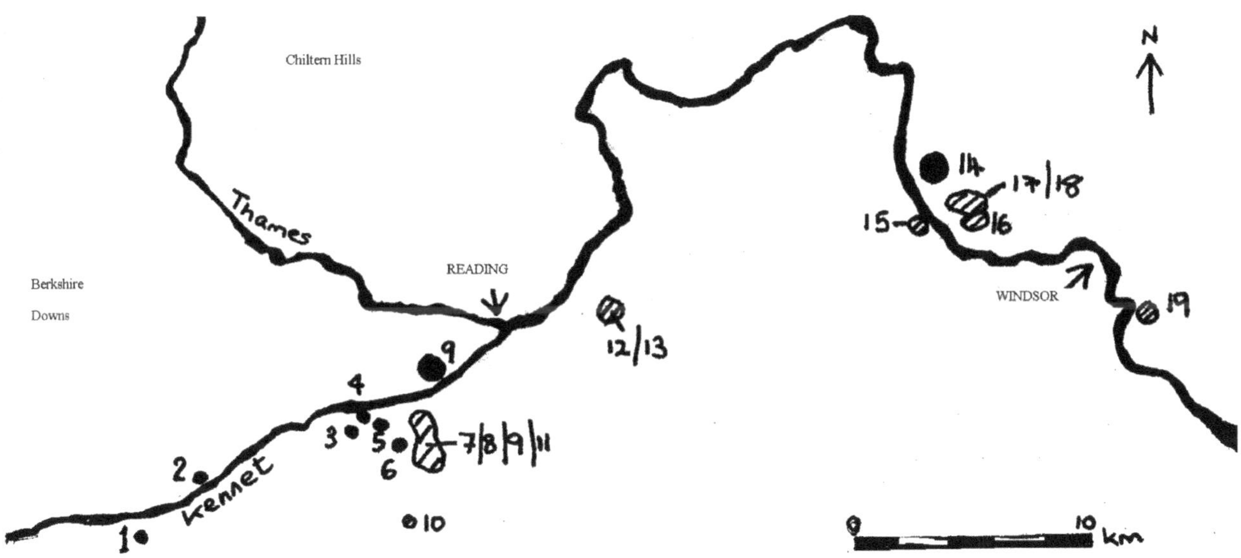

Figure 13: Sites and features in the Middle Thames Valley as recorded by Yates, 2007:34

Hatched areas represent field systems. Numbered sites: 1. Brimpton. 2. Aldermaston Wharf. 3. Field Farm, Burghfield. 4. Anslow's Cottages. 5. Knight's Farm 6. Pingewood. 7. Moore's Farm. 8. Reading Business Park. 9. Marshall's Hill (enclosure). 10. Grazeley. 11. Hartley Court Farm. 12. Land west of Park Lane, Charvil. 13. East Park Farm, Charvil. 14. Taplow. 15. Weir Bank Stud Farm. 16. Eton Rowing Lake. 17. Marsh Lane East. 18. Lake End Road. 19. Datchet. (List includes some sites where occupation doubtful).

SU 7069	Hartley Court Farm **(11)**
SU 7069	Reading Business Park Area 5 **(8)**
SU 7271	Marshall's Hill, Reading **(9)**
SU 7363	Riseley Farm, Swallowfield
SU 7674	Duffield House
SU 7775	East Park Farm, Charvil **(13)**
SU 7775	Land west of Park Lane, Charvil **(12)**

The Kennet- Reading cluster of sites is centred around the confluence of the rivers Kennet and Thames, mostly in the floodplain on small gravel islands, especially those in the Burghfield area. Most of the sites are within 1.5 km of each other. While some sites seem to be small, single-phase or even only seasonally occupied (such as Pingewood), others seem to have been extensive settlements with permanent occupation over some centuries. Together, they form a "notable density of occupation" in the Lower Kennet valley (Lobb & Rose, 1996:79).

The subsistence economy has been characterised as "exclusively pastoral" (ibid: 82) with the exception of traces interpreted as flax fields at Smallmead Farm; only the most westerly settlement, at Aldermaston Wharf, and Wickham's Field near Reading have evidence of both cereal growing and processing, and Aldermaston Wharf is the only settlement with grain storage facilities. Grain processing has been identified at Area 3100 at Reading Business Park (Moore & Jennings, 1992). The suggestion is that grain was imported. The site at Anslow's Cottages had evidence of a probable jetty in a paleo-channel of the Kennet, as well as areas of trampled earth caused by cattle (Butterworth & Lobb, 1992). Most sites had bones of cattle, sheep/goat and pigs, with small quantities of red and roe deer and occasionally, a small number of horse bones, but like almost all the sites except Field Farm have produced no grains of cereals at all. The series of excavations at Reading Business Park , adjacent to Smallmead Farm, have revealed an extensive area of occupation and utilisation throughout much of the Late Bronze Age. While much of the site is now landscaped and developed, it is still possible to gain an impression of the earlier landscape in neighbouring smaller parcels of land to the west of the development site on the same contour and alignment. In the valley bottomland are a number of very low raised gravel islands (less than a metre above flood level) ringed by alder and reeds, accessed by narrow causeways. Area 5, for example, is an island just 23-32cm higher than the floodplain level (Moore, 1992). The nature of this landscape suggests that line-of-sight links between the settlements would have been impossible, but the accoustic nature of the area would facilitate audible contact. Terraces rise up behind the settlement areas to the north relatively quickly; to the south, the valley is somewhat broader, but rises more steeply up towards modern Burghfield and the Grim's Ditch. The southern terraces would appear to offer a much better environment for settlement than the floodplain, being sheltered from prevailing winds, better drained, on reasonably good soils, within easy reach of the riverine resources, wide enough for use, and providing excellent viewscapes across the valley. It is therefore puzzling why the Late Bronze Age settlements are clustered on ground that would have been seasonally flooded,with a water table that today reaches up to 20cm below the surface of some of the gravel islands (and was probably worse during a period of climatic downturn), which are surrounded by a clayey alluvium (Moore & Jennings, 1992), and which attracts thick river mists at many times throughout the year. From the sites of the settlements themselves, the gentler rises to the north are less visible than those to the south and west (the beginning of the Berkshire Downs, the promontories of which gained a number of hillforts in the Middle Iron Age which are clearly sited to overlook the two valleys). If settlements had been placed on the southern terraces, however, they would have had no view south and very little west, and would have been easily approached from these directions. The actual siting of the settlements would offer warning of incursions from the south and west, and the nature of the access to the gravel islands would possibly have made aggressive approach difficult and restricted.

Numerous houses and other structures have been identified at Reading Business Park and are thought to be broadly contemporary - 20 houses in Area 5, a total of 12+ houses in Areas 3100 (Moore & Jennings, 1992) and 3000B (Brossler, 2001), and structures in all areas including 4- and 6- post buildings, fence lines, flax retting pits, pit alignments and ditches. There were seven or eight waterholes and a number of 'activity areas' of various types. Domestic crafts included bone working, spinning and weaving, and flax processing, and evidence of metal working including moulds for spearheads. Exotic finds included a shale bracelet. The evidence for the subsistence economy seems to centre on stock farming, with a possible cheese press among the finds, although the kill pattern for cattle suggests both dairy and beef farming were being practised. Weeds of arable cultivation were identified, but the small size of the associated fields has been associated by the excavators with the growing of flax (of which there is some pollen evidence) rather than wheat. Seeds of emmer and spelt wheat were found in Area 3100, along with a Lower Greensand saddle quern. Beans (*Vicia fabia)* are also represented (Moore & Jennings, 1992) .

The SMR entry for this site suggests that in Area 3100, the almost circular arrangement of the houses, and the presence of pollen associated with hedges, may suggest a hedged enclosure. A U-shaped ditched enclosure was excavated in Area 7000B, 88 m. long and tapering from 44 m. wide at its north end to 32 m. wide at the southern extremity.

Human remains have been found at this site in contexts which are not typical, although there is insufficient evidence to determine cause of death in any of the specimens. While small amounts of human bones have been found at a number of sites, here two crouched adult female burials and three adult cremations (possibly of only parts of the body) were found within the settlement area (Boyle, 1992:98). The site also produced a burnt mound, which had apparently seen repeated use (Moore & Jennings, 1992).

The site at Marshall's Hill, adjacent to an Early Bronze Age disc barrow, has been described as an "aggrandised settlement" on the basis of a record of its enclosure (Bradley, *pers.comm.* cited in Yates, 2007). It was excavated before the Second World War, and records are not sufficiently complete to be able to offer confirmation of this status, although its position on a relatively high spur on the north side of the valley suggests that it would have been in a good position to have had an oversight of many of the other settlements.

The Windsor-Maidenhead Sites

(numbers in brackets indicate sites identified on Fig. 14 below)

SU 9179	Taplow? (site destroyed without investigation) **(14)**
SU 8783	Furze Platt, Maidenhead
SU 8380	Hurley
SU 9480	Cippenham
SU 9578	Eton Wick
SU 9278	Dorney Reach

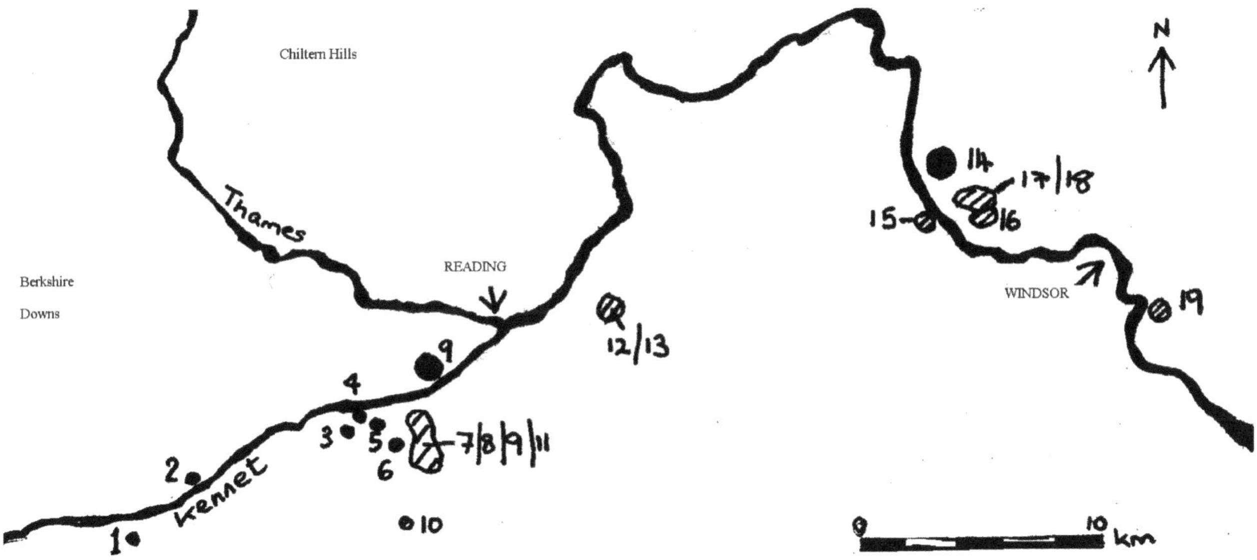

Figure 14: Sites and features in the Middle Thames Valley as recorded by Yates, 2007:34
Hatched areas represent field systems. Numbered sites: 1. Brimpton. 2. Aldermaston Wharf. 3. Field Farm, Burghfield. 4. Anslow's Cottages. 5. Knight's Farm 6. Pingewood. 7. Moore's Farm. 8. Reading Business Park. 9. Marshall's Hill (enclosure). 10. Grazeley. 11. Hartley Court Farm. 12. Land west of Park Lane, Charvil. 13. East Park Farm, Charvil. 14. Taplow. 15. Weir Bank Stud Farm. 16. Eton Rowing Lake. 17. Marsh Lane East. 18. Lake End Road. 19. Datchet. (List includes some sites where occupation doubtful).

Occupation seems to have ceased at nearly all settlements in this cluster by the 7th century BC, and if it continued into the Iron Age, it was not for long. As elsewhere in the study area, there is a notable shortage of Early-Middle Iron Age information.

The area between Windsor and Maidenhead has signs of extensive Middle Bronze Age co-axial field systems, some of which can be seen to take their alignments from, or to respect, earlier Bronze Age barrows; the settlement evidence, however, is more scanty at present. The site at Taplow has been described as a possible aggrandised centre, but the lack of investigation before destruction means that this cannot be confirmed. The area has produced a great deal of metalwork from the Thames and its tributaries, including many weapons (York, 2002), but less occupation material than in the other two areas.

The site at Cippenham consisted of an apparently single phase site with insubstantial buildings (Ford, 2003), while that at Eton Wick is identified by a single posthole and a truncated ditch; the Furze Platt site was recorded in advance of rapid gravel extraction work, so that only part of the site could be dealt with, but two roundhouse structures were identified, along with pits, wells, hearths and postholes. Again, this was a single phase site and was artefact-poor and has been described as of low status and with a poor economy, perhaps part of colonisation onto poorer soils in the 9th or 8th century BC (Berkshire SMR Number 00618.00.000).

It would appear that if a major settlement was associated with the large number of fine metal artefacts found in the area, this has yet to be found, unless the destroyed evidence at Taplow represents it. At present, therefore, while the general remarks which pertain to the other two clusters almost certainly apply to the Windsor-Maidenhead group, there is little that can be specifically discussed.

The Staines-Stanwell Sites

(numbers in brackets indicate sites identified on Figure 15 above)

TQ 0073	Manor Farm, Wraysbury
TQ 0074	Wayland's Nursery, Wraysbury
TQ 0168	Muckhatch Farm, Thorpe
TQ 0170	Thorpe Lea Nurseries I **(46)**
TQ 0170	Whitehall Lane/Milton Park Farm, Egham
TQ 0171	Petter's Sports Field, Egham
TQ 0171	Runnymede Bridge **(41)**
TQ 0267	St. Ann's Hill
TQ 0269	Thorpe
TQ 0271	The Close, Vicarage Road, Staines **(45)**
TQ 0371	2-8 High Street, Staines **(43)**
TQ 0372	Tilly's Lane, Staines (?) **(40)**
TQ 0373	Staines Moor
TQ 0373	Poyle, Stanwell **(36)**
TQ 0469	Fairylands Caravan Park **(47)**
TQ 0471	Leacroft Road, Staines
TQ 0475	Stanwell **(31)**
TQ 0578	Perry Oaks Sludge Works **(29)**
TQ 0565	Former Marconi site, Addlestone
TQ 0569	Home Farm, Laleham **(48)**
TQ 0570	Matthew Arnold School, Staines **(44)**
TQ 0572	Hengrove Farm, Staines
TQ 0574	N of Park Road, Stanwell
TQ 0577	Prospect Park, Harmondsworth **(21)**
TQ 0662	(+TQ 0563) Wey Manor Farm, Addlestone **(51)**

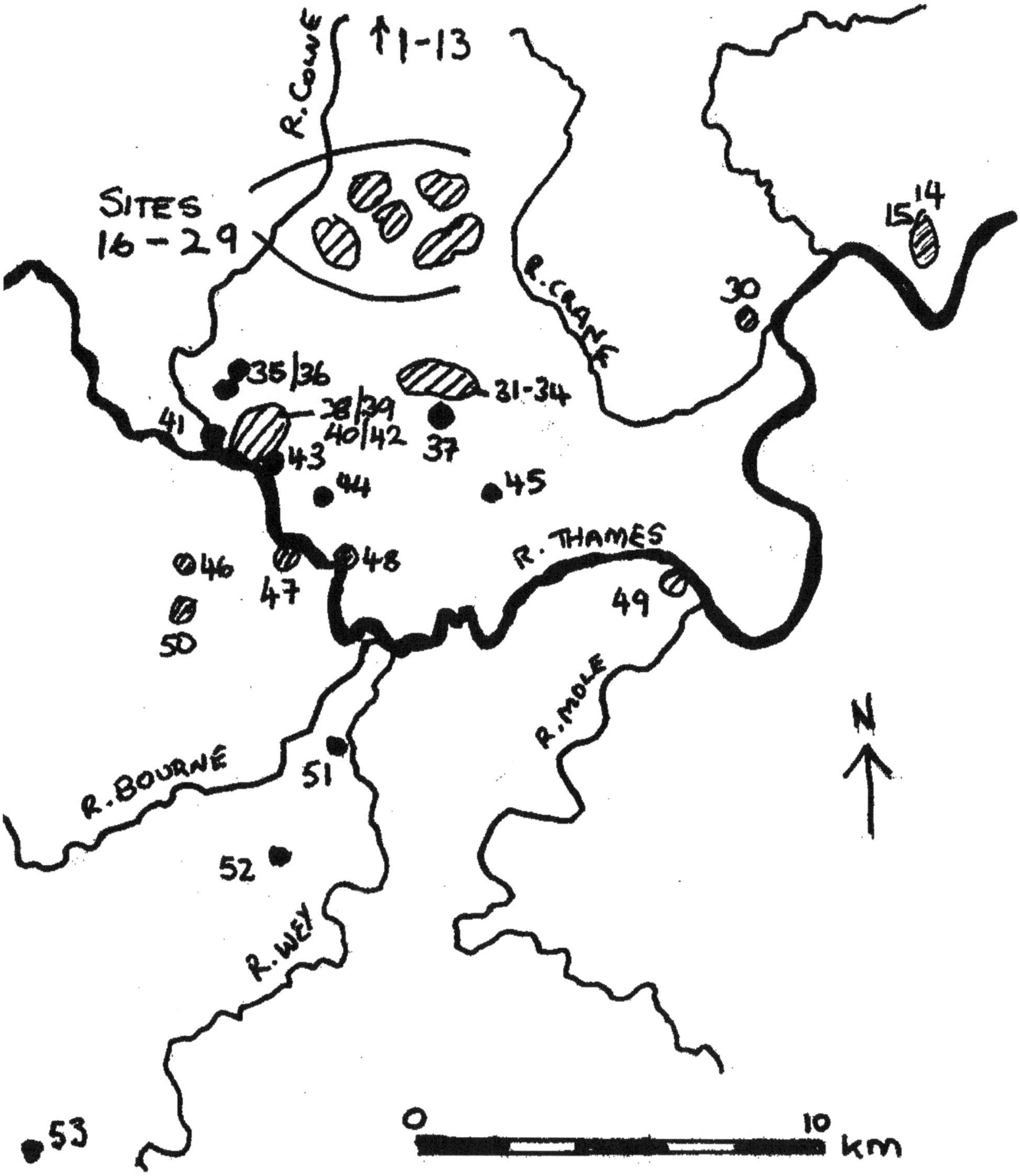

Figure 15: Sites and features on the West of London terraces recorded by Yates, 2007:33

Hatched areas represent field systems. 1-15 Sites outside of the present study area. 16. Wall Garden Farm. 17. Holloway Lane. 18. M4 widening/ gas main site. 19. Imperial College Sports Ground. 20. Home Farm, Harmondsworth. 21, Prospect Park. 22. Home Farm, BFI Quarries. 23. Nobel Drive, Heathrow. 24. Cranforld Lane, harlington. 25. Airport Gate. 26. Neptune Road, Heathrow. 27. Heathrow Northern Runway. 28. Heathrow Airport. 29. Perry Oaks Sludge Works. 30. Bankside Close, Isleworth. 31. Stanwell. 32. Cargo Point Development, Bedfont Road, Stanwell. 33. Heathrow Terminal 4. 34. Stanwell Road, East Bedfont. 35. Lower Mill Farm, Stanwell. 36. Poyle, Stanwell. 37. Mayfield Farm (enclosure). 38 and 39. Church Lammas, Staines. 40. Tilly's Lane, Staines. 41. Runnymede. 42. Staines Central Trading Estate. 43. 2-8 High Street Staines. 44. Matthew Arnold School. 45. Vicarage Road, Sunbury. 46. Thorpe Lea Nurseries. 47. Fairyland Caravan Park, Laleham. 48. Home Far, Laleham. 49. Hurst Park. 50. Junctions 12 and 15 (M25). 51. Wey Manor Farm, Addlestone. 52. Broadoaks Estate, W. Byfleet. 53. Whitmoor Common. (NB: A number of these sites fall outside the study area and are not discussed in the following).

TQ 0768 Staines Road Farm
TQ 0873 Mayfield Farm, Heathrow **(37)**
TQ 0876 Caesar's Camp, Heathrow
TQ 1468 Hurst Park, East Molesey **(49)**

This group of sites is spread over a large area of low-lying floodplain and terrace gravels and clays, in a landscape much dissected by river channels and marshland before historical and modern drainage, canal construction and reservoir building was undertaken. The number of sites increases yearly, as a result of large scale development, particularly around Heathrow Airport, and gravel extraction industries. Associated with the cluster of sites around Staines and Stanwell may be outlying groups of sites identified outside the study area around Uxbridge and as far south as Addlestone and Whitmoor Common (Yates, 2007). They lie within an area which has produced much evidence of Neolithic occupation (including a causewayed enclosure and a cursus monument), Early and Middle Bronze Age settlement, and Late Iron Age activity (including the temple site at Heathrow). Very little Early or Middle Iron Age evidence has been recovered, however, leading to an impression of abandonment and depopulation after about 700BC until about 400-300BC.

Actual evidence of houses is hard to establish, due to the nature of the soils and the amount of disturbance suffered on these flat, often flooded, deposits. Nor is environmental evidence without its problems, as pollen does not always survive particularly well, and the acid-rich gravels preserve bone material badly in many instances. However, a number of post-built structures have been identified, as well as ditches and pits. It is, however, very difficult to establish dating or phasing for structures on many of these locations.

The subsistence economy seems to have been increasingly focussed on stock-keeping (Yates, 2007). Faunal evidence has been recovered from some sites, most notably Runnymede Bridge (Needham, 1991), where the remains of cattle, sheep/goat, pigs, horses, and dogs have been found. At this site, a fast and 'dramatic' fall in the stature of the cattle has been noted in the earlier part of the Later Bronze Age, which is thought to represent an increase in the size of herds at that time (Serjeantson, *pers.comm.*). The kill pattern at Runnymede Bridge suggests a mix of beef and dairy farming. There is only very occasionally any evidence for exploitation of wild species (usually just a very few bones of red deer and some wild boar). Many other sites, however, had produced circumstantial evidence for stock farming, in the form of drove roads, watering holes and paddocks (for example, at Home Farm, Laleham (Hayman, 1998) and Perry Oaks (Barrett et al, 2001)). Runnymede Bridge has produced some evidence of wheat (*Triticum diococcum* and *triticum spelta*) as well as barley (*Hordeum sp.)*, rye *(Secale cereale)*, oats *(Avena sp.)* and fat-hen *(Chenopodium album)* (Needham, 1991), as well as some fragments of quernstones and two sickles at Petters Sports Field (Needham & Longley, 1980) but on the whole, evidence for cereal growing is sparse. One possible charred pea was also found at Runnymede. However, a palaeo-channel at the site has preserved pollen, which includes some weeds of cultivation as well as many weeds of pasture (Needham, 1991). The Heathrow terrace area was extensively organised for farming over some 150 sq.km. with most of the large scale Middle and Late Bronze Age co-axial field systems taking advantage of the terrace gravels which are, at this point, the widest area of farmland along the whole Thames corridor.(Yates, 2007:32-34), although it was not primarily directed towards the production of cereals (Yates, 2001). The shift towards a greater emphasis on stock farming is illustrated by the phased changes in the Perry Oaks system, where droveways, waterholes and a drafting gate were inserted into the pattern over time (Barrett et al, 2001; Maloney, 1999).

A few settlements in this area have been identified as unusual because of the presence of enclosures - Mayfield Farm, Thorpe Lea and Perry Oaks - or a waterfront structure - Runnymede Bridge. It has been suggested that these settlements represent elite or 'aggrandised' centres in a new settlement hierarchy (Yates, 2007; Needham & Longley, 1980). The presence of cooking pits at Perry Oaks (Barrett et al, 2001), and large deposits of smashed pottery at Thorpe Lea (Cotton, 2001) may represent traces of communal feasting (Yates, 2007), while at Runnymede, environmental evidence including types of plant remains not local to the site has suggested periodic visits to the site by non-locals, perhaps dependents or clients from a hinterland territory (Needham & Spence, 1996). Masses of other debris and the presence of 'burnt mounds' and spreads of burned flint may also suggest feasting activities at this site. Runnymede Bridge is situated on an 'ey' - a small gravel islet close to the river bank. Certainly up to medieval times, much of the channel of the Thames west of London was a complex series of channels and gravel 'eys', the latter being separated from the banks by small channels which may have been seasonally dry. They were popular sites for settlement in all periods (cf. Chertsey, Bermondsey) because the gravel offered drainage to form dry footings just above the flood levels. In later centuries, some of these islets were particularly selected to house industrial processes, particularly pottery-making, as they had water to hand but were sufficiently remote that they avoided spreading fire to other buildings, as was the case at Kingston upon Thames. Some evidence for industrial processes has been found at Runnymede, including perforated clay slabs, moulds and other equipment for pottery manufacture and metal working (Needham, 1991), and metalworking debris has also been found at the nearby Petters Sports Field site amongst the contents of a large Late Bronze Age hoard (O'Connell, 1986).

Many of the excavated sites in this list have produced evidence for textile-working (spindle whorls, loom weights), suggesting either sheep farming on the upper terraces or importation of wool from communities occupying these areas, and artefacts of worked bone and antler. Flintwork is, on the whole, poor and scrappy, often consisting merely of utilised struck river pebbles. Other stone artefacts, including quermstones, have been recovered, mostly formed from sandstones from the Greensand Beds lying between the North Downs and the Weald, or occurring on the sandy terraces on the northwest Surrey borders with Hampshire and Berkshire. (Freestone, 1991).

The waterfront structure at Runnymede Bridge, variously interpreted as a palisade or a jetty, may represent a feature that contributed to the growth and wealth of communities in this area - the control of long-distance trade, especially that with mainland Europe (perhaps including new sources of metal as mentioned above). Artefacts of European stylistic origin, such as brooches and pins, have been found at Runnymede Bridge, Petters Sports Field and reported among other finds in or near the Thames in this region; a catalogue of these items is not yet available, but they appear to be relatively more numerous in this area than further upriver. Amber and shale artefacts also appear at Runnymede Bridge. The excavators are of the opinion that control of river traffic was a major reason for the siting of this settlement and its palisade, which they believe that while potentially defensive, was primarily directed towards economic gain. Boats could have been used to control passing traffic "for maintaining political advantages and exacting economic fringe benefits....or by.... enhancing commerce through the provision of facilities such as storage, basic supplies, safe moorings, hospitality... and perhaps a ready market for exchange" (Needham & Longley, 1980:421).

Apart from the river finds and the Petters hoard, other hoards have come to light at Laleham and just to the east of the study area at Syon Park, Kingston Hill and Teddington. Axes, spears and swords figure commonly among the objects that make up these hoards and among the river finds discussed by York (2002). It is noticeable that the further the area downstream on the Thames, the greater is the number of river depositions of metalwork - between Molesey and Runnymede, covering the Staines-Stanwell site cluster, 51 items are recorded (with a further 102 in the next ten miles downstream); between Windsor and Maidenhead there are 64 items, and only about 12 in the Reading-Kennet zone (York, ibid:82). A correlation has been noted between metal finds and the areas of better soil along the valley (Needham & Longley, 1980).

The pottery found on the Late Bronze Age sites presents problems. The forms and fabrics, while having some relationship to Middle Bronze Age styles derived from the Deverel-Rimbury tradition of Wessex, are generally particular to the Thames valley and to settlements along the tributaries, particularly to the south. This 'post-Deverel-Rimbury' material and succeeding 'Iron A' forms are not securely dated, and different reports seem to ascribe different dates to similar material. Doubts have been cast particularly on the ceramics from the latest phases of several sites - in some cases these have been interpreted as early appearance of Early Iron Age forms, and in others, as an indication that an Iron Age date might be a misinterpretation for these wares. This is a debate which the author of this paper does not feel qualified to enter; where dating of sites depends on the dates from the ceramic assemblages, therefore, an area of doubt pertains. What is clear is that the Late Bronze Age ceramics of the Thames valley have become part of a different production and distribution system from those found in the Middle Bronze Age. Given the uncertainties of dating, the general conclusion seems to be that most, if not all, of the settlement sites had gone out of use by the end of the 7th century BC, with one or two possibly surviving into the early 6th. In most instances, there is a notable gap in the monument and artefact records for the area until the Late Iron Age or Romano-British period. The Stanwell field systems were abandoned some time in the Late Bronze Age, and no Iron Age material has been found there (O'Connell, 1990), and at Heathrow Terminal 4 a waterhole which had been in use for centuries silted up and went out of use around the same time (MoLAS, 1996). At Perry Oaks, the field ditches silted up. There was almost no Iron Age material in the primary fills, and no signs of any agricultural development in the area after the end of the Bronze Age (Barrett et al, 2001). It appears that the whole regional economic system failed, but the reasons for this are unknown.

Presence of Causal Correlates

There seems to be an increasing body of evidence to suggest that a global climatic downturn occurred in the Late Bronze Age. Many of the sources of this evidence have been collated by Franzén and Larsson (1998), who carried out studies in Sweden and Tunisia: "Both studies showed that climatic downturns happened around 1500BC and 1000BC, involving general cooling, extreme amounts of precipitation, followed by flooding. From many other sources of information it is obvious that these events were sudden and occurred world-wide...In the latter period, dendrochronological data...points to one low peak of tree growth at 1159BC and another one at a few years after 950BC". (1998:148). They also cite evidence from Assyrian texts, Alaskan environmental data, Lapland glacier expansion and low tree limits, Polish records of violent increases in water levels in rivers and lakes, Southern Swedish studies of lake levels which were higher after c1000-900BC, Northern Irish evidence of increase in lake levels c1000-600BC with a significant high stand at Loughnashade around 950BC and higher levels elsewhere, and Lake Van, Eastern Turkey, which rose almost 80m between 971-902BC (varve dating). Further studies based on pollen evidence are recorded by Dark (2006).

Lake rises are reported for the Great Salt Lake, USA, Waldsea Basin, Canada, Lake Titicaca, and Lago Cardiel in Western Argentina, and there was a peak in sedimentation

Figure 16: Conjectural reconstructions of the waterfront structure at Runnymede Bridge (after Needham, 1991:115)

in Midge Lake, Livingstone Island (Antarctica) culminating around 900 BC. There is also a significant break in bog oak growth from Ireland and France around 950BC, apparently a widescale change, not merely dependent on local conditions, and peat growth evidence from Sweden shows significant increase in the late second/early 1st millennium BC. The annual growth rate doubled around 900BC (ibid: 150).

In central Europe water levels in lakes reached levels far over normal (in the Saalachsee, Ammersee, Federsee, Bodensee and others), resulting in the build up of new high beach terraces and widespread abandonment of lake pile dwellings. New lakes formed near München, Tölz, Memmingen and Ravensburg. All this occurred soon after 1000 BC (first recorded by Gams & Nordhagen, 1923). Other studies have suggested that the Late Bronze Age saw the beginning of an unusual rise in precipitation (Lamb et al, 1966) and temperature changes resulting in an overall decline of crop ripening levels of about 150m between the 12th and 7th centuries BC (Parry, 1978). At Steng Moss, Northumberland, peat growth grew from 0.65mm/year in the period 1320-820BC to 4.47 mm/year in the years 832-765BC and this is attributed to catastrophic climate decline (Barber, 1982). This, it is argued, caused general contraction of settlement, abandonment of marginal lands, general retreat from the uplands, and far-reaching dislocation of settlement and agricultural systems. (It should be noted that most experts recognise that there can be a hiatus of a few hundred years between the causal event and the observable changes in peat formation). Indications of new crops, more intensive farming methods and food collection and storage methods, with possibly severe population decline are suggested by Burgess (1985).

Research in the Ripple Brook catchment in central England involved C^{14} dating of floodplain deposits - dramatic increases in sediment deposition were observed for the Late Bronze Age/Early Iron Age (900-300BC)(Brown, 1985). The authors believe this was possibly due to human activity - deforestation and cultivation of catchment slopes and resultant soil erosion. (Brown & Barber, 1985). The sediment increase reached around 500%. Annual erosion values of 20-30 tons/km^2 prior to 950BC suddenly increased to values of around 140 tons/km^2. It is hard to accept that such change could be wholly based on changes in agricultural practice, and with the other evidence cited above, climatic change must surely be implicated. Dark (2006) believes that there is little evidence to support large scale agricultural abandonment, particularly any which was caused by erosion or depletion of soils. Her studies seem to indicate that, although there is very little available data from southern central England, this region shows a picture of arable abandonment (though probably not pasture desertion) which is more likely to be related. in the river valleys and low-lying areas, to a change in the water table caused possibly by increased precipitation. She notes that the areas apparently affected by abandonment seem to be those which were brought under cultivation earlier than in other regions, and perhaps these regions were organised in such an optimal fashion for agriculture that farmers found it less easy to adapt to environmental changes. However, she suggests that other mechanisms were responsible for changing farming in the region; changes in water tables cannot explain all the other forms of change that are recorded archaeologically.

The suggested causes for these climatic events vary. Franzén and Larsson suggest that the presence of glass spherules, increased marine calcium levels and tektites in their studies are possibly related to comet activity and suggest that there may have been an asteroid impact in the western North Atlantic, probably on a shelf off North Africa or Southern Europe between 1000-950BC (1998:157), a possibility supported by Baillie (1998: 111-112) based on data from Irish tree rings and Greenland ice cores showing-significant events around 1628-23 BC and 1159-41BC, with a major temperature trough around 1100BC, also matched by Fennoscandian tree ring data, while van Geel et al (1998) have noted a sharp rise in the C^{14} content of the atmosphere around 850BC contemporaneous with an abrupt climate change from relatively warm and continental to cooler and wetter conditions - the transition of Sub-boreal to Sub-atlantic. In the Netherlands low lying areas were abandoned and there was a rise in the water table, and extension of fens and bogs. This C^{14} oscillation, they believe, was caused by changes in solar activity. Increased cosmic ray flux caused increased cloudiness and precipitation, amplifying the effects of reduced solar activity (1998:162). The different forms of analysis and the variable nature of the evidence for these events result in an area of uncertainty for precise dating; in some instances there may be a notable delay between cause and discernable effect accumulation. However, overall the evidence suggests that there is likely to have been an appreciable change in climate between the 12th and 10th centuries BC.

Whatever the causes of the climate deterioration, farming strategies would have been forced to adapt to poorer conditions. Higher arable fields may have been abandoned because of increasing failure of crops to ripen over shorter, wetter summer seasons; valley bottoms would have been flooded seasonally or become too boggy for settlement. In such circumstances, an increasing reliance on stock farming may have been a preferred response, allowing utilisation of the wet valley floors as water meadows and the unproductive hilltops as summer pasture. Settlement might be expected to relocate to sheltered terraces above seasonal floods or perhaps to utilise waterfront gravel islands, such as those identified at Wallingford and Runnymede on the Thames. Excavations at Staines (Crouch & Shanks, 1984) have suggested that a gradual rise in the water table restricted occupation to one central gravel island during the Later Bronze Age and much of the Iron Age, effectively along the present High Street. A bank described as a flood defence work (Yates, 2007) seems to have been constructed at Tilly's Lane, Staines. The prevalence of stock raising in the Late Bronze Age has been recognised. "Pastoralism was not just the predominant aim: there is evidence that it became paramount by the LBA" (Yates, 1999:163).

The Middle Thames region is likely to have experienced effects including longer, colder winters, higher levels of precipitation, localised flooding, cooler summers and a curtailed growing season.

Changes to subsistence patterns on the Wessex uplands have been noted; pre-existing 'Celtic' fields appear to have been overlain with long linear ditch systems usually thought to relate to a widescale adoption of stock-farming or 'ranching' (discussed by, for example, by Cunliffe, 2004; Bowen, 1978). In the Thames Valley bottom and terraces, dairying may have provided an alternative, utilising meadows seasonally flooded and thus likely to have enjoyed early and enhanced fodder growth. The various sites considered in the study area show comparatively little evidence for cereal growing, suggesting a reduced reliance on grain and possibly a change in dietary habit. However, low-lying flood meadows would be unsuitable for over-wintering of stock; this may have been carried out on the better drained terraces above the settlements.

The size of the population could possibly be deduced from the relative acreage of field systems, and from the number and size of the settlements. The degree of urbanisation, the tree cover, and the proximity of Heathrow reduce the availability of such data. Settlements, however, are known from excavation; they appear to occur in groups, with stretches of land between each that seem to have very little evidence of habitation.

The Kennet-Reading group of settlements comprises at least two dozen sites; some of these may be sections of the same larger settlement, whilst others appear to be contemporary but distinct, even if only physically separated by a few hundred metres. Not all have produced evidence for houses, but all have some trace of domestic occupation. Dating is imprecise at many of these sites, and there is some dispute over the major dating tool, pottery, with some estimates of stylistic date being rather later than others, but by and large, most of these sites seem to have been active between the tenth or ninth centuries to the eighth or early seventh centuries BC. Any estimates of the size of population of these settlements must necessarily be tentative, but it may be possible to suggest that an overall figure centred in the middle of the occupation period might be in the range of 500 people, on the basis of a conservative estimate of an average of 4-5 families per site.

Estimates for the population of the next most easterly group of settlements, around Windsor and Maidenhead, are even less secure. The sites recorded are nearly all known from trial trenching only, and minimal information has been recovered.

	No. Structures	Estimated population
SU 6067 Aldermaston Wharf	2	8-10
SU 6169 Ufton Nervet		
SU 6266 Raven Hill		
SU 6270 NE of Bostock Lane, Englefield		
SU 6669 Green Park	15	60-75
SU 6770 Field Farm, Burghfield	?	?
SU 6770 Field Farm II		
SU 6770 Field Farm 3		
SU 6769 Amner's Farm		
SU 6769 Wickham's Field (M4 Services)	?	?
SU 6769 Field Farm MSA		
SU 6869 Pingewood Area C		
SU 6869 Moore's Farm, Pingewood		
SU 6869 Knight's Farm Subsite 4		
SU 6870 Knight's Farm Subsite 3		
SU 6969 Pingewood, Burghfield	2	8-10
SU 6971 Anslow's Cottages, Burghfield	?	8-10
SU 7069 Smallmead Farm	30	88-110
SU 7069 Hartley Court Farm	?	?
SU 7069 Reading Business Park Area 5	20	80-100
SU 7271 Marshall's Hill, Reading		
SU 7363 Riseley Farm, Swallowfield	0	?
SU 7674 Duffield House	?	?
SU 7775 East Park Farm, Charvil	1	4-5
SU 7775 Land west of Park Lane, Charvil	?	?

**Population estimates are based on proportion of buildings thought to be contemporary on each site at any one time; sites with no buildings recorded are included on the basis of other recorded occupation debris.*

Table 7: Estimates of population per site
Kennet-Reading Group

	No. Structures	Estimated population
SU 8380 Hurley		
SU 8783 Furze Platt, Maidenhead	2	8-10
SU 9179 ?Taplow	--	--
SU 9278 Dorney Reach	2	8-10
SU 9480 Cippenham	1	4-5
SU 9578 Eton Wick	?	?

**Population estimates are based on proportion of buildings thought to be contemporary on each site at any one time; sites with no buildings recorded are included on the basis of other recorded occupation debris.*

Table 8: Estimates of population per site
Windsor-Maidenhead Group

	No. Structures	Estimated population*
TQ 0073Manor Farm, Wraysbury	?	
TQ 0074Wayland's Nursery, Wraysbury	?	4-5
TQ 0168 Muckhatch Farm, Thorpe		
TQ 0170 Thorpe Lea Nurseries I	2	8-10
TQ 0170 Whitehall Lane/Milton Park Farm, Egham		
TQ 0171 Petter's Sports Field, Egham	6	24-30
TQ 0171 Runnymede Bridge	(?)	up to 50?
TQ 0267 St. Ann's Hill		
TQ 0269 Thorpe		
TQ 0271 The Close, Vicarage Road, Staines		
TQ 0371 2-8 High Street, Staines		
TQ 0372 Tilly's Lane, Staines	(?)	
TQ 0373 Staines Moor		
TQ 0373 Poyle, Stanwell		
TQ 0373 Staines Moor		
TQ 0469 Fairylands Caravan Park		
TQ 0471 Leacroft Road, Staines		
TQ 0475 Stanwell	2?	8-10
TQ 0565 Former Marconi site, Addlestone		
TQ 0569 Home Farm, Laleham		
TQ 0570 Matthew Arnold School, Staines		
TQ 0572 Hengrove Farm, Staines	1	4-5
TQ 0574 N of Park Road, Stanwell		
TQ 0577 Prospect Park, Harmondsworth	1+	8-10
TQ 0578 Perry Oaks Sludge Works		
TQ 0662 (+TQ 0563)Wey Manor Farm, Addlestone	1	4-5
TQ 0768 Staines Road Farm		
TQ 0873 Mayfield Farm, Heathrow		
TQ 0876 Caesar's Camp, Heathrow	11	36-40
TQ 1468 Hurst Park, East Molesey	3*	

* Population estimates are based on proportion of buildings thought to be contemporary on each site at any one time; sites with no buildings recorded are included on the basis of other recorded occupation debris.

Table 9: Estimates of population per site
Staines-Stanwell Group

The most easterly group of settlements is the Staines/ Stanwell area; figures for some sites clearly indicate the difficulties of excavation in advance of road construction or in urban areas, as there is comparatively little evidence for houses, but a very large number of pits, gullies, ditches, middens and other occupation features have been recorded in the vicinity of most of these locations. Bearing this in mind, it seems likely that the overall population might have been not dissimilar to the Kennet-Reading group in a similar landscape of riverside terrace and gravel island locations within the floodplain of the Colne, Thames and Duke of Northumberland's rivers.

Working on an absolute minimum, assuming just one family per site at which no evidence of structures has been recovered, the populations of each area in the Late Bronze Age may therefore be estimated at:

Kennet-Reading Group *328-410*

Windsor-Maidenhead Group *30-35*

Staines-Stanwell Group *210-245.*

Actual populations may have been much higher; in the Staines-Stanwell area, the majority of sites have been identified as a result of small-scale development, rather than the large acreage available in some of the Kennet-Reading sites, and few of the sites are recorded as producing any form of settlement boundary, so it can be assumed that in some cases at least, larger occupation areas remain to be discovered at a number of locations (notably Staines High Street, where a number of very small excavations along some 200+m of the street have produced some data, recorded on the Surrey SMR). Further find spots and antiquarian observations in the Windsor-Maidenhead area hint at more extensive occupation, particularly around Bray (Ford, 1987).

To establish whether or not there has been a significant rise in population in the Late Bronze Age, similar assessments need to be made of Middle Bronze Age settlements. This is problematic, as there are very few known sites in the valley itself, and indeed, it has been suggested that a form of construction and settlement may have been in use during the period which leaves minimal archaeological recovery potential (Farwell et al, 1999:67-8).

However, several occupation sites have been identified in the Staines-Stanwell area - at SU 9965 on the 60m contour, at Mixnam's Pit, Thorpe at 15m OD (TQ 0469), at Beomand's Farm, Chertsey (Abbey Meads) (TQ 0468) also at 15m OD, and predating the Late Bronze Age settlements at Home Farm, Laleham, The Close, Staines and Hengrove Farm, Staines. Many other occupation sites have been noted, but there is insufficient data for precise dating within the Bronze Age in the SMR records or published notes. A number of cemeteries of the Middle Bronze Age have been identified, most sited on or near the plateaux edges just above the valleys. Middle Bronze Age barrows are recorded at SU 9965, and urnfields are reported at TQ 0970 and at Shepperton Studios (TQ 0768 at 15m OD) but the relevant data has been lost. Three barrow cemeteries are recorded in the East Berkshire area - at Ascot (TQ9168), Cockmarsh (TQ8889) and Rapsley Farm (TQ8864), with a number of isolated barrows and dubious mounds. Overall, evidence for population in the area in the Middle Bronze Age suggests a relatively sparse occupation density.

The abandonment of sites on higher ground at the end of the Middle Bronze Age (around 1000BC) is likely to have been at least partly motivated by the climatic changes. The more exposed heights would have suffered from harder frosts and more severe winter conditions, although they probably would have been spared the worst of any flooding. Soil depletion is unlikely to have been a factor in the region; indeed, the weather conditions are likely to have improved fertility locally by bringing nutrient rich silts down river, except where flooding was worst. Other natural threats, such as erosion, are not generally a problem on these geologies except in periods of drought.

The possibility of epidemic disease cannot be discounted as a causal factor for change; murrains of stock, crop plagues and human infectious diseases might have occurred as a result of the deteriorating weather conditions, as occurred during the first half of the fourteenth century AD. However, there is no conclusive evidence for any such plagues in the study area in the later Bronze Age.

Change as a result of developments in social complexity is a possibility in the period. By the later Iron Age, Roman commentators claim the existence of defined tribal groupings in Southern Britain, a state of affairs borne out by evidence from coinage and pottery distributions. The process of Iron Age tribe formation is not clearly understood, and the point of origin of these tribes is unknown. The development of Early Iron Age hillforts might suggest that the process was under way by the sixth or fifth centuries BC; the identification of what have been called 'elite centres' in the Late Bronze Age (at, for example, Ram's Hill, Marshall's Hill, Queen Mary's Hospital, Carshalton and perhaps Mayfield Farm, Heathrow) could possibly indicate that distinct chiefdoms were beginning to emerge at this early date, perhaps part of the formative process of tribal organisation. Barrett (1980) cites Hodder (1977) who suggests that social identity finds expression through stylistic unity and distinction of artefact types, and that this becomes particularly evident at times of stress, both internal and external through resource competition. This can, perhaps, be discerned in several aspects of the Later Bronze Age in the Middle Thames region.

Associated with the development of elite centres and Early Iron Age hillforts may be some evidence of a developing territoriality along the Middle Thames. This section of the valley appears as a distinctly different territorial entity to that of the region to its west by the 8th century BC "with a sharp cut-off along the edge of the chalk overlooking the Thames Valley and conform[ing] to the line of the Berkshire Grims Ditch" (Barrett, 1980:92). Barrett also notes the differences in settlement forms and siting between the two areas, with clustering of settlements in the valley and a dispersed pattern on the uplands. The apparent clustering of settlements has been noted above; a similar clustering of field systems has also been noted (Yates, 1999). Yates dates the laying out of these field systems to the later Middle Bronze Age and Late Bronze Age; they are concentrated in discrete groups including a Runnymede-Petters group, and a Marshall's Hill, Reading group, and differ noticeably from agricultural systems further west. He associates these groups of organised, co-axial fields with Late Bronze Age elite settlements and with a growing emphasis on stock-raising. The distribution of settlement concentrations, metalwork finds and pottery distribution appears to match that of the groups of field systems.

Finally, there do seem to be hints of changes in access to resources in the Late Bronze Age in the Thames Valley. Metal 'founders' hoards appear in several locations - on Kingston Hill, at Petters Sports Field, at Mixnam's Pit, Thorpe and at Hythe, west of Staines, perhaps suggesting an increased requirement for recycling of bronze.

There are also apparent changes in the distribution of pottery and other artefact types. Barrett (1980) has identified a marked boundary between pottery style zones in the Middle Thames Valley and those north of the Ridgeway; this boundary also occurs along the line of Grim's Ditch, just to the west of the outliers of the Kennet-Reading group of settlements. Metal working activity and quantity is much greater in the valley area than the uplands, while evidence for flint working is generally more scanty and small scale in the valley (Barret & Bradley, 1980b).

Presence of Preparatory Correlates

Fortifications are commonly regarded as prime evidence in support of the occurrence of warfare. In the Late Bronze Age, there is very little evidence for such constructions in Southern Britain. Within the study area, there are only a few sites which hint at any level of fortification. Sites possibly identified as elite settlements include Marshall's Hill (Bradley, 1986) and Mayfield Farm, both thought to have been enclosed by a bank and ditch, and at Runnymede Bridge evidence for a palisade along the waterfront has been discovered (Needham & Longley, 1980). Further enclosed sites have been excavated at Ram's Hill and Carshalton (Barrett & Bradley, 1980). Possible enclosed sites noted on the Surrey SMR also include the former Marconi site at Addlestone, Petters Sports Field, Thorpe Lea Nurseries and Wey Manor Farm, but at none of the latter was excavation extensive enough to be able to confirm this potential.

East Berkshire has a number of short lengths of linear earthworks, ranging between 70m and 500m in length. which have been tentatively dated to the Late Bronze Age, but only one of these (Maidenhead Thicket - TQ857808) has been excavated (Berkshire SMR). Nevertheless, one account suggests these might bear comparison with the linear earthworks of the chalk uplands to the west (Ford, 1987).

The proliferation of weaponry and armour is another of the most commonly accepted forms of evidence for the occurrence of warfare. The Late Bronze Age does appear to offer confirmation in this respect. Finds of swords, spearheads, chapes and ferrules are relatively common in the study area, particularly from the river Thames or its tributaries, often poorly provenanced as a result of recovery during dredging operations or museum deposition as stray finds. The history and methodology of dredging operations and acquisition of metalwork finds is discussed briefly by York (2002).

As has often been observed, the Thames has always functioned as both a highway and as a natural boundary, and has liminal attributes (e.g. Lambrick, 1998). Of 302 artefacts in collections that are identifiable as coming from the Thames or its tributaries, as far east as the Hogsmill River, Kingston-upon-Thames, 46% date from the Late Bronze Age. These may be separated into several groups, relating to 20-mile stretches of the valley:

West (Reading to Medmenham) approx. 10 artefacts

Central (Medmenham to the R. Colne) approx. 48 artefacts

East (R.Colne to the R. Hogsmill) approx. 65 artefacts.

The majority of finds of this period are weapons or weapon accoutrements such as ferrules and chapes. Weapons occur in other contexts in the region; the assemblage from the possible founder's hoard at Petters Sports Field, Egham, contained three Ewart Park hilt fragments and two Carp's Tongue hilt fragments. There were seven blade fragments of the earlier type, nine of the latter, and one of unidentified form. There were also four spearhead fragments, of which two were conjoining, a chape and a ferrule. It is possible that some of these weapons were made at the nearby site of Runnymede Bridge (based on analysis of metal composition and the presence of casting moulds) (Needham, 1986).

A third common source of evidence for warfare is in the form of iconography. There seems to be none of this in any form in the period in this region. Settlement abandonment is to be seen, however. Abandonment, or at least cessation of activity, of Middle Bronze Age sites on higher ground may have been due to climatic change, but the threat of offensive action cannot be altogether discounted.

Whether or not the communities of the abandoned settlements were the same as those of the valley bottom settlements of later date, the choice of the valley bottom settlements could be interpreted as defensible. Many sites are placed on very low gravel islands within in the floodplain, accessible for much of the year only by narrow causeways. Some of these must have been fairly uncomfortable places to live, cold and damp, with a water table no more than a few centimetres below the land surface. Indeed, at Tilly's Lane, Staines (TQ 0371), structures have been identified as Bronze Age flood defence measures (Yates, 2007). However, any aggressor would have been faced with an unpleasant and slow struggle across muddy marshland, and at certain times, the villages would have been concealed by mist and fog. In addition, the close proximity of the settlement sites could have provided additional security if mutual defensive reaction was agreed. In both the Kennet-Reading and Staines-Stanwell groups the majority of the occupation sites are less than one kilometre apart, many probably within sight of each other if there was little forest, and most of which could have been visible together from vantage points on the upper terraces. A signal such as a horn blast or a column of smoke or fire could have alerted neighbours and allowed them to mass for defence, as was the practice among the Hopi and Aricopa in Arizona (LeBlanc & Rice, 2001).

NGR	SITE NAME	HEIGHT (average)
Kennet-Reading Group		
SU 6067	Aldermaston Wharf	55m OD
SU 6169	Ufton Nervet	53m OD
SU 6266	Raven Hill	90m OD
SU 6270	NE of Bostock Lane, Englefield	52m OD
SU 6669	Green Park	38m OD*
SU 6770	Field Farm, Burghfield	42m OD*
SU 6770	Field Farm II	42m OD*
SU 6770	Field Farm 3	42m OD*
SU 6769	Amner's Farm	48m OD*
SU 6769	Wickham's Field (M4 Services)	48m OD*
SU 6769	Field Farm MSA	48m OD*
SU 6869	Pingewood Area C	40m OD*

SU 6869 Moore's Farm, Pingewood	40m OD*
SU 6869 Knight's Farm Subsite 4	40m OD*
SU 6870 Knight's Farm Subsite 3	41m OD*
SU 6969 Pingewood, Burghfield	40m OD*
SU 6971 Anslow's Cottages, Burghfield	39m OD*
SU 7069 Smallmead Farm	38m OD*
SU 7069 Hartley Court Farm	38m OD*
SU 7069 Reading Business Park Area 5	38m OD*
SU 7271 Marshall's Hill, Reading	60m OD
SU 7363 Riseley Farm, Swallowfield	45m OD
SU 7674 Duffield House	54m OD
SU 7775 East Park Farm, Charvil	49m OD
SU 7775 Land west of Park Lane, Charvil	49m OD
*settlement agglomeration	
Windsor-Maidenhead Group	
SU 8380 Hurley	76m OD
SU 8783 Furze Platt, Maidenhead	44m OD
SU 9179 ? Taplow	22m OD
SU 9278 Dorney Reach	21m OD
SU 9480 Cippenham	24m OD
SU 9578 Eton Wick	30m OD
Staines-Stanwell Group	
TQ 0073 Manor Farm, Wraysbury	18m OD
TQ 0074 Wayland's Nursery, Wraysbury	17m OD
TQ 0168 Muckhatch Farm, Thorpe	15m OD
TQ 0170 Thorpe Lea Nurseries I	15m OD
TQ 0170 Whitehall Lane/Milton Park Farm, Egham	15m OD
TQ 0171 Petter's Sports Field, Egham	18m OD
TQ 0171 Runnymede Bridge	12m OD
TQ 0267 St. Ann's Hill	45-60m OD
TQ 0269 Thorpe	15m OD
TQ 0271 The Close, Vicarage Road, Staines	14m OD
TQ 0371 2-8 High Street, Staines	15m OD
TQ 0372 Tilly's Lane, Staines	14m OD
TQ 0373 Staines Moor	20m OD
TQ 0373 Poyle, Stanwell	18m OD
TQ 0469 Fairylands Caravan Park	14m OD
TQ 0471 Leacroft Road, Staines	14m OD
TQ 0475 Stanwell	15m OD
TQ 0565 Former Marconi site, Addlestone	18m OD
TQ 0569 Home Farm, Laleham	13m OD
TQ 0570 Matthew Arnold School, Staines	19m OD
TQ 0572 Hengrove Farm, Staines	15m OD
TQ 0574 N of Park Road, Stanwell	20m OD
TQ 0577 Prospect Park, Harmondsworth	23m OD
TQ 0578 Perry Oaks Sludge Works	20m OD
TQ 0662 (+TQ 0563) Wey Manor Farm, Addlestone	16m OD
TQ 0768 Staines Road Farm	10m OD
TQ 0873 Mayfield Farm, Heathrow	20m OD
TQ 0876 Caesar's Camp, Heathrow	24m OD
TQ 1468 Hurst Park, East Molesey	8m OD

Table 10: Later Bronze Age settlement/occupation sites in the Middle Thames valley – heights above datum

Thus the change in settlement location from the higher ground to the valley bottom might possibly have been a deliberate choice in the face of the threat of aggressive action. Furthermore, the agglomeration of settlements within a restricted radius might also have been a deliberate defence strategy to enable massing defence. There is little evidence for any internal defensive morphology in those settlements excavated so it seems unlikely that any kind of static defence was visualised. It is difficult to ascertain whether or not guard settlements were placed on the edge of the settlement agglomerations, as we cannot be sure that all the component sites have been yet identified, but consideration of this possibility and the likely locations of such sites might form future research agendas. It is perhaps worth noting that, within the limitations of the present data, there is a pattern of outlying settlements associated with the Reading group and with the Staines group that is concentrated to the south and west of the major agglomerations, with few to the north. It might therefore be possible to postulate a guard role for these settlements; conversely, their more scattered nature might suggest that they existed in a region of comparative safety. However, the majority of the possible enclosed settlements are those to the south and west, and thus the possibility exists that it was from these directions threats might be expected, and an early warning role for these sites cannot be discarded without further investigation.

Line-of-sight links are also hard to determine because of a number of factors, including a dearth of pollen information about contemporary tree cover, and possibly changes in the height of occupation levels due to alluvial effects. However, a look-out placed on the river terraces would probably have been able to see a considerable proportion of the component settlements in a cluster in the Reading-Kennet group, and have been capable of giving the alarm. Audible signals could have been very effective as the relationship between the terraces and valley floor in this area offer good acoustic properties. The wider floodplain of the Thames-Colne group tends to offer fewer opportunities for oversight of the area, except on Cooper's Hill, above Runnymede.

Between the settlement clusters, there are areas which, so far as current knowledge extends, are lacking habitation or have only very small, scattered hamlets or farms. In fact, the concentration of settlement is very noticeable - the sites listed in this study represent over 90% of the Late Bronze Age sites in the general area that are recorded on the county SMRs. In the boroughs of Runnymede and Spelthorne, all but two Late Bronze Age sites fall within the study area.

If these intermediate areas really are largely unoccupied, (although not necessarily unused), they may represent no-go areas between communities, formed as a peace-keeping strategy. They might otherwise represent contested areas between emerging chiefdoms. Thirdly, they might represent areas considered unsafe for habitation, subject to raids and ambushes. A single farmstead and its family could be highly vulnerable, as could workers in the fields if in small numbers. It might be considered much safer to move house to a larger settlement, and to go to work in the fields in larger groups, possibly accompanied by a protective cohort of armed warriors.

The field systems along the Thames appear in the Middle Bronze Age, and do not survive after the Late Bronze

Age (Yates, 1999). Most evidence seems to be for small plots for flax growing and for paddocks and pastures. The move towards a greater reliance on stock keeping has been discussed above; the scarcity of cereal remains in the area is also notable, and Lobb & Rose (1996) suggest that cereals consumed on the sites in the Kennet/Thames confluence area may have been imported from drier settlement areas further west. These changes may have been climate-driven, and may also be part of a general shift towards stock farming in southern England in the period, linked in the Wessex uplands to the linear or 'ranch' boundaries which have been recorded succeeding arable field systems (McOmish, Field & Brown, 2002). It is possible that this change is also related to changes in social structure, with herd numbers being related to the importance or power of emergent groups or individuals. The slighting of earlier boundaries during this process has been likened to the highland clearances (Field, 2001). Another activity that might be associated with the Thames Valley settlements is textile production; in addition to evidence for flax fields and retting pits at Reading Business Park (Yates, 1999; Moore & Jennings, 1992), most sites have produced spindle whorls and loomweights, and Yates (1999:163) suggests that this might have been a major activity.

Trading relationships in the Later Bronze Age have undergone a notable shift; the previously dominant areas of Wessex have far less metalwork and exotic goods than the erstwhile peripheral Thames Valley. The settlements of the Thames Valley became part of a Europe-wide exchange network, gaining in wealth as a result.

> *"The collapse of the Wessex Culture had led these groups on diverging paths, the hierarchy in the Thames Valley developing through control of prestigious artefacts as well as land, whilst social differences in Wessex still depended more directly on the control of territory and of an agricultural surplus."*
>
> (Bradley, 1980:69).

It is not clear that any unoccupied zones developed; however, sparsely occupied zones are visible between the large clusters of settlement, and this sparseness includes a lack of observed field systems and artefact finds (Yates, 2007, 1999). The reasons for this could be many - the presence of a very few independently-minded families, a system of control of large hinterlands managed by the leaders of the clusters, or a deliberate decision to avoid disputes between clusters by leaving land unclaimed at territorial margins. Some of the relatively empty land is poor - sandy, thin soils; much is currently forested. Nevertheless, this land was in use in the Middle Bronze Age, and field systems have been recorded in parts of these areas (Yates, 1991). While soil exhaustion may have played a part in some areas, others are much richer - the area to the south of the Staines cluster is fertile alluvium, but has produced very little evidence of any Late Bronze Age occupation.

Presence of Functional Correlates

There does not appear to be any evidence in the study area for burned or deliberately destroyed sites, nor is there any unequivocal evidence of skeletal trauma resulting from weapons injuries, in either human or animal bone recovered from sites in the region. The lack of such evidence may result from the apparent change of burial rite from around 1000BC which results in an almost total scarcity of skeletal evidence, as well as, perhaps, the effects of flood scouring and soil acidity.

However, there is some evidence of weapons use in the form of physical damage to swords and spearheads found in hoards and other contexts. Weapons finds in the region include large proportions of damaged and destroyed items. In a study of these artefacts, it was found that 82% of spearheads had damage associated with use, half of which had been later destroyed (York, 2002). Types of damage noted included broken tips, edge chipping, crushing damage and percussion breaks. Destruction levels are seen to rise in the Late Bronze Age (from the Wilburton phase onwards) to 60%, and the researcher has suggested that this may indicate a rise in conflict and a greater need to disable spearheads (York, 2002). The condition of recovered spear ferrules is generally too poor for detailed assessment.

In the case of swords, 84% overall showed evidence of use and 59% showed evidence of deliberate destruction. Destruction methods include bending and breaking off of tips, bending (possibly with the aid of heat) until snapping occurred, and breaking by striking across the blade. Some swords were broken by these means into several pieces. Deliberate destruction of swords appears to become more prevalent as one travels eastwards along the river. As with spearheads, destruction occurs more often in the Late Bronze Age, peaking in the Ewart Park phase. Particularly high levels of destruction were noted at Staines and Bray, and lower levels at Weybridge and Windsor (ibid).

The meanings of these depositions are unclear, but could belong to one or more of a number of activities. A strong possibility is an association with a new form of funerary ritual; some weapons show traces of burning which may connect them to cremation rites (cf. the large number of 'burnt mounds' recognised along the Thames and its tributaries). Votive offering may also explain the practice, with or separate from mortuary rites. These could have been combined with ostentatious destruction of valuable weapons in a form of competitive consumption display.

> *"Another possibility is staking a claim to a territory, or making a statement about ownership and control...a kind of surrogate warfare. Deposition*

could also represent triumphal display after defeating an enemy"

(York, 2002:90).

In addition to the deliberately inflicted damage, there is evidence of damage resulting from usage:

	Deliberate damage LBA (MBA)	***Use damage LBA (MBA)***
Spearheads	60% (33%)	88% (75%)
Swords	68% (39%)	84% (80-83%)

Use-wear damage was also noted on finds from the Petters hoard, including edge-sharpening on both spearheads and swords. Various indentations and scrapes were also considered to represent use-damage, as were indications of edge damage. It has been noted that damage from deliberate attempts to destroy the weapon tends to occur face-on, rather than edge-on (Needham, 1990:26).

Presence of Resultative Correlates

At the end of the Bronze Age, there are major changes in the region under consideration. Most of the settlements, apparently after hundreds of years of occupation, are abandoned by the end of the 8th century BC, and none show much habitation after the end of the 7th century (Yates, 2001, 1999: Lobb & Rose, 1996). Thus the area as a whole underwent change in many of the factors indicated by the list of correlates: subsistence, trade, development of unoccupied zones, and population movement.

"In the Middle Thames there is evidence of economic and social dislocation in the Late Bronze Age. On the Heathrow terraces, settlements and fields were abandoned during this period. This pattern of upheaval is also apparent at Burghfield [Kennet-Reading cluster] where much of the area was abandoned.....No new field systems were created in the Early Iron Age in the Thames Valley, nor on the surrounding uplands of the Berkshire Downs or Marlborough Downs"

(Yates, 1999:167).

Settlement location for the succeeding period appears to have been away from the river valleys, as little trace has been found for subsequent habitations until at least the Middle Iron Age, even in areas subjected to intense archaeological investigation. Later settlements are to be found above the 50m contour level, but for the most part are smaller and, whilst relatively densely scattered, are slightly further apart from each other. The site at TQ868842, almost uniquely, has produced evidence for Early Iron Age occupation in the form of two hut circles and associated pottery, but it stands alone in its region (ibid).

The trade in bronze seems to have collapsed completely, and with its fall, the communities of the Thames Valley ceased to exist as organised, powerful groups. Bradley (1980) discusses the possible mechanisms for the end of the bronze trade, and notes Hodson's (1971) observation that Southern Britain became isolated from Europe in the Hallstatt D period. The developing use of iron in Europe may have led to a drying up of the supply on which the prestige of the Thames Valley settlements was built. The absence of succeeding settlement information means that we also lack information about subsistence and trade in the ensuing period.

No evidence for trophies, memorials or other forms of iconography has been found. There is little or no information available about the health, sex ratio or standard of living of people in the Early Iron Age in the region from skeletal data; burial information is almost completely absent.

It is not possible, with the data currently available, to say whether the field systems associated with the Late Bronze Age settlements were equally abandoned wholesale. Certainly, the ditches and enclosures close to the settlements which have been excavated do go out of use, but the lack of close association with more distant possible agricultural territory means that the same determination cannot be made across the region. Again, changes in biodiversity within the abandoned settlement areas can be observed, but no data on biodiversity within the wider region has been identified during this study.

Those settlements which do survive into the Early Iron Age do not display significant changes in artefact types. The pottery sequences are unclear; local post-Deverel Rimbury assemblages are poorly dated, and may extend well into the Iron Age period.

There is only a very small amount of burial evidence for the Late Bronze Age sites, and that appears to be atypical. The presence of burnt mounds and signs of feasting may point towards some form of ritual activity (Yates, 2007; Barrett et al, 2001; Cotton, 2001), and the visits of outsiders at Runnymede may also be interpreted in this light (Needham & Spence, 1996). It is possible, though unproven, that the mortuary ritual consisted of riverside cremations, accompanied by deposition of the ashes and of votive offerings of tools and weapons into the water.

The rise of hillforts in the region is mainly evident from the Middle Iron Age, although Caesar's Camp (SU863657) has produced chance finds of Early Iron Age haematite-coated pottery (Surrey SMR). There are relatively few such sites, compared with the chalklands, and those which are found are sited at some distance from the river terraces, on chalk or gravel spurs, which nevertheless offer good view sheds over stretches of the valley. There is no clear relationship between the hillforts and the Later Bronze Age settlement distribution. Indeed, likely sites of Late

Bronze Age occupation in the region do not develop into hillforts at all (such as Queen Mary's Hospital, Carshalton, Ram's Hill, Marshall's Hill or Kingston Hill), and so far, the known hillforts to the south such as St. Ann's and St. George's have produced little evidence of settlement before their construction in the Iron Age. To the north, hillforts are not found until the summit of the scarp slope of the Chilterns. Late Bronze Age occupation is indicated at Ivinghoe Beacon and at Tottenhoe, but these sites are oriented towards the Oxford Basin rather than the London Basin, and are physically remote from the study area.

The appearance of these sites may possibly be taken as an indication of the development of a more formalised territoriality in the region, that eventually comes to be included in the tribal identity of the Atrebates, with Calleva as an oppidum. The problem of where the Thames-side populations went after the abandonment of the Late Bronze Age/Early Iron Age settlements is unsolved. Needham (1980) has speculated that settlements might be found along the spring-line of the North Downs to the south of the valley, but so far little such evidence has been recovered. Later Iron Age settlement, however, is well-attested at a number of different locations, including Brooklands (Weybridge), Percy Gardens (New Malden), Alpine Avenue (Tolworth), Hawk's Hill (Fetcham) and Bocketts Farm (Leatherhead). The first three of these settlements are all sited on rising ground beside tributaries of the Thames, while the last two overlook the Mole Gap. Field systems have been associated with Bocketts Farm, noted on air photographs (John Hampton, pers. comm.). This small sample of Later Iron Age sites seems to indicate a move away from the Thames, perhaps associated with a shift in economic emphasis towards a hinterland with exploitation of a range of downland, slope and riverine resources, or just possibly, a retreat from what had become a contested frontier between the tribes to the north (Trinovantes and Catuvellauni) and the Atrebates, a border now guarded by developed hillforts.

Case Study II - Gallia Belgica

Description

To offer a comparison to the prehistoric case studies, where there is no written or epigrammatic evidence for warfare, this case study will collate evidence appropriate to the list of correlates which derives from the period of the mid-late first century BC in Gallia Belgica. This is the period which contains Julius Caesar's conquest of Gaul, and his record, along with those of other Roman writers, offers an historical account of actual warfare in a defined territory at the very end of prehistory. The region has been the subject of a variety of archaeological research, which, although patchy and incomplete, provide a number of local and regional studies related to these events. Gallia Belgica is a very large area, whose boundaries were altered for administrative reasons on several occasions during the period of Roman occupation. For the purposes of this study, the area discussed will be restricted to the lands of the tribes mentioned by Caesar that lie north and east of the river Seine as far as the river Rhine, bounded on the north by the Channel coast and on the south by the wooded difficult terrain of the Ardennes *(figure 20)*.

The region is crossed by a number of important rivers - the Seine, Somme, Oise, Marne, Aisne, Sambre, Scheldt and Rhine, with smaller tributaries. To the west, the region is made up of chalk geology with gently rounded hills and plateaux in Picardie, Artois and the Pas-de-Calais; further east the flat loess and alluvial sandy plain of Flanders and Hainault was bounded by coastal marshes to the north, and rising rolling land to the south (across much of Brabant) before reaching the convoluted and relatively inhospitable gorges, bogs and heights of the Ardennes (composed of a variety of harder, mostly igneous rocks including granite).

The tribes living in this region in the mid first century BC are called by Caesar "Belgae", a general and apparently flexible term, under which he from time to time includes some German groups and even some possibly Norman or Armorican tribes. The tribes of the Belgic heartland, and those most usually included in this description, are, according to Caesar, the Suessiones, the Viromandui, the Ambiani, the Atrebates, the Bellovaci, the Remi, the Nervii, the Eburones, the Menapii and the Morini. The Treveri are also sometimes included. It seems clear that these tribes had very varied stages of political development at the time of Caesar, and while there were some alliances and clientage systems, these seem to have been fluid; there are relatively few hillforts, and little sign of the development of any proto-urban oppida of the kind found further south in Gaul.

The region seems generally less wealthy than its more southerly neighbours – the rich graves of the Marne and Champagne areas of the Late La Tène are not paralleled across Gallia Belgica except for a few examples in the Ardennes (Wightman, 1985).

Subsistence activities are not well represented in the archaeological record, but what information exists seems to suggest that arable agriculture was most dominant in the southeast of the region (on the *limon* soils of the edge of the Paris Basin) and in Picardie, where large fields have been identified by aerial reconnaissance (Agache, 1979). Smaller 'Celtic' fields have been identified elsewhere, particularly in the Netherlands (Van den Broeke, 2005). Cattle and pigs were apparently the dominant forms of domestic stock in the northern plains, with sheep and goats more common in the southwest (Meniel, 1997). The coastal tribes may have been able to extract and trade in marine salt. The area was generally mineral-poor in the period, though with some iron ore in the Ardennes and coal deposits may have been exploited along the Sambre-Meuse corridor.

Information from aerial survey, notwithstanding the work of Roger Agache and others, is by no means comprehensive. There are a number of causes for this, beyond the degree of archaeological research undertaken and surveys published. Parts of the region are heavily wooded; other regions may

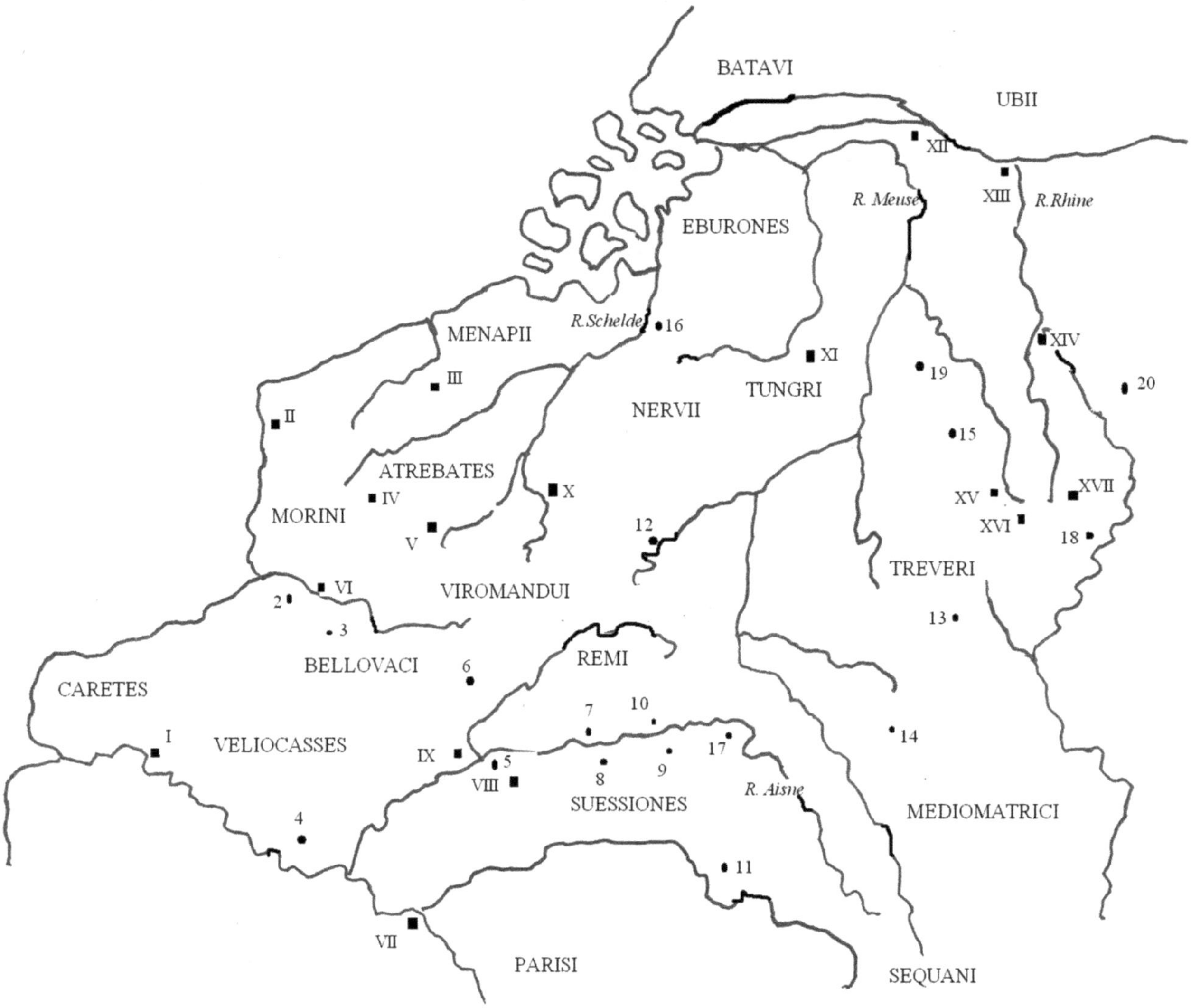

Figure 17: Tribes, with some major hillforts, and civitas capitals in Gallia Belgica

Civitas Capitals		*Hillforts*			
I	*Rouen (Rotomagus)*	*1*	*Incheville*	*18*	*Martberg*
II	*Boulogne (Gesoriacum)*	*2*	*Liercourt-Hérondelle*	*19*	*Mont Lassois*
III	*Cassel (Castellum)*	*3*	*La Chaussée-Tirancourt*		
IV	*Thérouanne (Tarvenna)*	*4*	*Mont César*		
V	*Arras (Nemetacum)*	*5*	*St. Pierre-en-Chastres*		
VI	*Amiens (Sarmobriva)*	*6*	*Gournay-sur-Aronde*		
VII	*Paris (Lutetia)*	*7*	*Pommiers*		
VIII	*Reims (Durocortum)*	*8*	*Villeneuve-St. Germain*		
IX	*Soissons (Augusta)*	*9*	*Vieux-Reims*		
X	*Bavay (Bagacum)*	*10*	*Vieux-Laon*		
XI	*Tongres (Atuatuca)*	*11*	*Vieux-Châlons*		
XII	*Nijmegen (Noviomagus)*	*12*	*Avesnelles*		
XIII	*Xanten (Colonia Ulpia)*	*13*	*Tetelbierg*		
XIV	*Cologne (Colonia Claudia)*	*14*	*Wallendorf*		
XV	*Trier (Augusta)*	*15*	*Otzenhausen*		
XVI	*Neumagen (Noviomagus)*	*16*	*Asse*		
XVII	*Wederath (Belginum)*	*17*	*Condé-sur-Suippe*		

have been subjected to heavy flooding in more recent periods. Much of the northern part of the region is very low-lying; climatic evidence suggests that at the time of Caesar's campaigns it was becoming drier (Haselgrove 1990), but in the post-Roman period and in the medieval era and post-medieval times, climatic deterioration has led to periodic inundations and sea ingress.

More generally, much of the region (particularly along the Somme and into western Flanders) has been substantially affected by the effects of the First World War and other conflicts - entrenchments, shell craters, cemeteries, military installations and other recent earthworks tend to obscure and confuse the archaeological picture and make excavation sometimes perilous. Certain areas have seen heavy industrialisation - around Lille and Armentières, and in the Charleroi area; others have been developed for ports (Calais, Boulogne, Antwerp) and as urban centres. Commercial and industrial developments have, however, enabled some large scale modern investigations, particularly around growing towns (such as Amiens, Arras and Tongres) and along communication corridors including high-speed rail links, motorways and canals between Paris and the Channel ports, for which full publication is pending (Haselgrove,2007).

An attempt to compare the correlates for war suggested in this study to this region in the respect of the historically documented Roman conquest which began in the 50s BC must take into consideration the varying state of archaeological research in the several countries and *départements* concerned, as well as the sometimes ambiguous and often patchy written record. The latter must be taken with some reservations, being specifically a Roman account written from the perspective of the victors and with the prejudices and assumptions of the time. Much of what occurred, where it happened and the causes for and results of events remain outside history. However, the writings of Caesar and others offer some framework against which the archaeological evidence can be compared and analysed. Interpretations of the events of the Caesarian conquest vary, but the main sequence is presented here as recorded in *De bello Gallico* and as commented upon by Wightman (1985).

In 57BC the Remi sent envoys to Caesar, reporting a massing of Belgic tribes to their north and offering the Romans a base of operations in their territory (around the Aisne Valley). Caesar marched north, crossing the Aisne not far from modern Reims, probably at Condé-sur-Aisne, and met an army of Belgae (apparently mainly Suessiones), inflicting heavy casualties among them. The Bellovaci (*clientes* of the pro-Roman Aedui) surrendered without a fight, their war leaders having fled to Britain. Caesar then turned north-east to face a combined force of Atrebates, Viromandui and Nervii, some 15-20 km into Nervii lands. The Nervii were apparently almost wiped out in the ensuing battle, although some non-combatants had reportedly retreated north into estuaries and islands, and their surrender was accepted. The Atuatuci, on their way to aid the other Belgae, turned back on hearing of the defeat and concentrated their forces into a single stronghold - either Namur or Mont Falhize. near Huy, on the Meuse. Using siege equipment, the Romans stormed this site and killed tens of thousands of people, selling over 50,000 survivors into slavery. A short campaign in 56 BC against the Morini and Menapii on the coast ended inconclusively, and a legion was left to garrison that area through the winter.

In 55BC the Usipetes and Tencteri crossed the Rhine and ravaged the lands of the Menapii and other tribes. Caesar took action against them at a battle on a river, either the Rhine itself or the Mosel. Much of the rest of that summer was spent in preparing for and undertaking the first expedition to Britain, using either Wissant-Sangatte or Boulogne as embarkation points (Wightman, 1985). The second expedition to Britain in 54BC was apparently cut short when news reached Caesar of widespread revolts among the Belgic tribes. The Treveri was particularly involved in these, as well as the Eburones and Nervii. This was a year of drought and hardship for the people of the region [evidence confirmed by tree-ring analysis (Hollstein, 1980)]. At this time some seven legions were operating in the region, with Caesar's headquarters being at Amiens. The Belgae employed Roman siege tactics against the invaders at Atuatuca. Roman forces countered this attack as well as subduing the Menapii (using causeways and bridges to reach their marsh strongholds), marching in strength into the Treveri territory and across the Rhine, and occupying sections of the lower Scheldt and middle Meuse valleys. There was widespread devastation across the region.

In 53BC Caesar launched another attack on the Nervii and the Eburones, resulting in the flight of many people into the Ardennes, the northern marshes or abroad. The fortress at Atuatuca was occupied by the Romans, who used it as a base to ravage the territory. The whole of the Eburones' territory was reportedly plundered, with the burning of villages and farms, and the destruction of crops and herds. Germanic peoples crossed the Rhine to join in the plunder, but also attacked some Roman units on or near the Sambre. Caesar withdrew his headquarters to Durocortorum, in Remi territory.

The following year saw the rebellion led by Vercingetorix, culminating with the siege at Alesia. Belgic tribes sent troops to try to lift the siege, including 2,000 Bellovaci, 5,000 each from the Suessiones, Ambiani, Mediomatrici, Petracorii, Nervii, Morini, and Nitiobroges and 4,000 from the Atrebates. The defeat of Vercingetorix was followed by 'mopping-up' operations in Belgic territory, with the capture of the chiefs of the Treveri and raids against the Eburones. However, by late 51BC and 50BC, Caesar seems to have been making conciliatory gestures towards local aristocracies, and there followed a period of calm. A further revolt of the Bellovaci occurred in 46BC, Agrippa campaigned in the region in 39BC, a further campaign against the Morini took place in 30BC, and in 29BC the Treveri were defeated by Nonius Gallus. Further problems are noted in 24BC and between 19 and 17BC, on both occasions apparently arising from Germanic incursions. Thus the period of recorded conflicts covers some seven years (57-51BC) of major campaigning, with a further series of sporadic fighting over the next 35 or so years - two generations of people.

The presence of correlates in this particular case is complicated by several factors, mostly due to the rise and influences of the Roman state. In any consideration of conditions of prehistoric warfare in North West Europe, it is evident that the appearance of this form of state polity did not occur, and thus much of the discussion must be tempered by the special political circumstances. Nevertheless, the events and effects of the period as recorded and as recoverable archaeologically represent the closest temporal occurrence of war to those of later prehistory and occur, at least on the part of the Gauls, in a prehistoric environment, and as such offer some basis for comparison and analysis.

Presence of causal correlates

The first causal correlate suggested is that of climatic deterioration during the period. This does not seem to have been the case during the mid to late first century BC. In fact, the evidence suggests a general amelioration of conditions, with drying of coastal marshes and an overall warming (Wightman, 1985; Haselgrove, 1990b; Krausse, 2003). However, the second correlate, that of demographic rise, does seem to be indicated. The number and size of settlements recorded in the La Tene phases C and D increases (Haselgrove, 1990a, 1990b, 1996a), and Bayard (1996) suggests that in Picardy there are settlements every three kilometres or so dating to the later La Tène, based on evidence from excavations along the lines of new autoroutes and the TGV. This rise is traced from the third century BC and seems to have been accompanied by agricultural intensification, although it is not clear which is cause and which is effect. Similar increases in settlement and cemetery numbers are recorded in Belgium from the start of the first century BC (Krausse, 2003). Wightman has calculated that, given a recorded figure of Belgic warriors and their allies in the rebellion of 57BC of 306,000, with a ratio of 1:5 or 1:4 warriors to non-combatants and using British estimates of pre-Roman Conquest population density of up to 11 people per km^2, the population of Gallia Belgica at the time of the Gallic War was approximately 1-1 ½ million people (1985:33). It may be incautious, however, to accept the historical figures, which may have been inflated for propaganda purposes, and the population total could well have been significantly lower.

Other subsistence and social threats can also be identified. Caesar and other Roman writers record the invasions of the region in previous periods by groups of Germanic settlers, including the various tribes called 'Cisrhenani' and the Atuatuci of the Tongres area, who were reputedly descendants of 6,000 wandering Teutones (Wightman, 1985). No dating is available to confirm these raids, but it may be possible to suggest some such incursions around the turn of the first century BC, based on evidence of hillfort usage, discussed below. Further German raids and population movements took place in the wake of the Caesarian campaigns, suggesting that this was an ongoing state of affairs. These movements may, at least in part, be responsible for the major settlement changes observable in the half century before Caesar, which will be discussed in more detail later. The effects of Roman intervention and trade in southern and central Gaul probably also had a destabilising effect, as raiding for slaves and booty was encouraged by the profits to be made in trading for Mediterranean products (Haselgrove, 1996a).

The period also seems to have shown increases in social complexity, although the degree of change varies from tribe to tribe, with those furthest away from central Gaul being less developed than those tribes which adjoined that area. Evidence comes in the form of the appearance of planned fortified settlements such as Condé-sur-Suippe and Villeneuve-St.-Germain (Haselgrove, 1996a:252), Pommiers, and the Titelberg (Haselgrove, 1987). Haselgrove describes a

> *"consistent trend, one of increasingly political centralisation the closer we move towards the zone of state formation in Central Gaul. It is only in the extreme north and east, in the coastal areas inhabited by the Nervii, the Germani Cisrhenani and other population groups, and across the Rhine, that we encounter a dispersed settlement pattern and other attributes of genuinely acephalous communities" (1987:111).*

Roymans (1990) has identified a number of different types of settlement, with scattered unfortified farmsteads dominating the northeastern part of the region (Northern Belgium, the Netherlands and the Rhineland), enclosed scattered settlements (*fermes indigènes*) more commonly found in Northern France, and fortified settlements of various sizes becoming smaller and less common generally from south to north. The French settlements display more orderly plans.

> *"In the [northern] lowlands the settlements had a loose internal arrangement and their locations were frequently shifted, particularly in the sand and peat areas. There is little evidence for status differences between settlements. In Northwest France better structured and more stable settlements existed.... There is evidence for vertical status differentiation among these.."*
>
> (Roymans, 1990:190).

Status evidence includes the degrees of wealth and trade contacts exhibited in the grave goods found in the associated cemeteries.

A further indication of political development is the appearance of coinage in the region, albeit in less numbers and with more limited metal issues than in more developed parts of Gaul. Indications suggest that coinage in Gallia Belgica can be first traced to the peoples of the Somme Valley, spreading a little later into the lands of the Suessiones, Nervii, Treveri and Morinii (Wightman, 1985). The coin evidence also helps to delineate the formalisation of territories, and indeed of political alliances, with links evident between the Nervii and Morinii, and then the

Treveri, and a different distribution marking the Suessiones and their contacts, which may have included the Ambiani and Remi (ibid).

The arrival of an aggressive Roman trading campaign in Central Gaul, which inevitably spread its effects into the tribes of the north east, certainly altered trading patterns and access to certain resources. The region remained firmly based in its own agricultural subsistence pattern, but the demand for Mediterranean luxury goods certainly seems to have permeated the aristocracy of many tribes, and may also represent an explanation for the rise of certain of the fortified centres, particularly those situated where they could command trade routes such as those of the valleys of the Aisne and the Somme. However, in some areas, according to Caesar and other Roman sources, the lure of the luxury goods market was resisted, because the people recognised its debilitating effects on Gaulish society, and this is borne out archaeologically by the scarcity of Roman imports in some areas, notably in those areas with Germanic populations (Hedeager, 1997; Hamilton, 1996) and in Picardy and other parts of the north and east (Bayard, 1996). Dressel 1 amphorae are well represented in the Aisne Valley and the area to the south, but extremely rare elsewhere in Gallia Belgica (Haselgrove, 1990a). Overall, therefore, we can say that in all probability, the availability of resources was generally unchanged, but in some parts of the region, the desire of the rising aristocracy for luxury items, and the means of acquisition of these perhaps through increased raiding in the territory of their neighbours, could have resulted in localised change to resource access and to alterations in patterns of trade for scarcer resources such as metal ores.

The real cause of the wars of the 50s BC is not identifiable archaeologically; there can be no physical evidence for the ambitions of a single individual, and that this individual comes from a far distant expansionist state also precludes traces of the development of invasion and conquest from being localised in the region in question. The problem of individual agency in archaeological cause and effect is apparently insuperable.

Presence of preparatory correlates

The presence of fortified centres has already been noted. The rise of hillforts and oppida in Gallia Belgica, however, predates the Roman war, and can perhaps be associated not only with changes in trade relations (at least partly caused by Roman mercantile expansion) but also with earlier conflict situations. In Hainault, fortification began as early as 450BC, although in at least one example (Le 'Bois du Grand Bon Dieu' at Thuin) repairs and additions were made immediately before the Roman conquest (Bonenfant,1998). Debord (1995a) dates Pommiers to between 130-120BC, while suggesting that Villeneuve-St.-Germain may actually post-date the conquest (to some time between 50 and 15BC) (ibid:1995b). Haselgrove (1996a) gives a slightly earlier date for Villeneuve (80-55BC) and a date for Condé-sur-Suippe of c.120-80BC. Wallendorf is dated to the later second or early first century BC, while Martberg and the Titelberg are both considered by Krausse (1999) to date from the latter half of the first century BC. It would appear that in some cases at least, we "can envisage a period of instability and successive crises stretching back to the Germanic invasions of half a century earlier (*De Bello Gallico II,4)* and subsequently fuelled by the political ambitions of rulers.., necessitating the construction of fortified sites on various separate occasions" (Haselgrove, 1990b:258).

The different sizes of hillforts and their distribution require brief comment. Just one of the Belgian hillforts is over 20 ha. in enclosed area, the Trier region has five over 20 ha., and the Rhineland none at all compared to 18 in the French part of the region. Where archaeological evidence is available from the interiors, there is a suggestion that the smaller forts were centres of storage and metal production, with small local elite occupation, while the larger excavated forts seem to have fairly large empty areas inside (Haselgrove, 2002); Caesar used both the word *oppidum* and the word *castellum* in describing the fortified centres of the region, and may have been recognising these differences. He noted that the Aduatuci abandoned the smaller sites to concentrate their forces in one oppidum when his troops reached their territory (DBG 2.29.2). A similar choice was made by the Suessiones, who, with 12 oppida in their territory, retreated *en masse* to just one (possibly Pommiers) (DBG 2.4.8), and the Remi concentrated their forces in one of their several oppida when facing attacks from other Belgic tribes (DBG 2.6.1).

Thus the implication is that the empty areas of the larger fortifications were intended to contain refugees and their stock during large-scale assaults. "This strategy reflects an existence of a certain hierarchy among the oppida within a civitas" according to Roymans (1990:201), although this hierarchy cannot, as yet, be related to evidence for density of population at these sites in normal circumstances.

The same comments regarding previous conflict appear to apply to the weaponry known from the period. The sanctuary site of Ribemont-sur-Ancre has produced an enormous mass of weaponry and armour dating from La Tène C1 to the beginning of D2, with lesser amounts of material, including Roman coins and amphorae from the later parts of D2, possibly post-Conquest (Brunaux, 2000). The percentages show a greater utilisation of the site in LTC1b:

LTC1a	around 10%
LTC1b	a little under 80%
LTC2	around 7%
LTD1	about 4%
LTD2	under 2%.

The material deposited at Gournay-sur-Aronde is nearly all (90%) datable to LTC, with no material ascribed to LTD (ibid.). Thus no proliferation of weaponry appears to accompany the period of the Roman conquest. However, the figures above are derived from votive deposits, and it may be posited that following the events set in train by Julius Caesar, the opportunity or motivation to make such deposits was lacking. It may also be supposed that much *materiel* would have been destroyed by the Roman forces, to prevent rebellion, or carried off as war booty by individual soldiers, a practice which still continues (*cf. The Times,* 26th December 2005 on the looting of Saddam Hussein's gold-plated weapons).

Iconography is also lacking for the period, although once again, the sanctuaries noted above can be taken to represent a physical iconography of war in previous periods. Wellington (2001) suggests that a survey of 'temple' groups in Northern France offers a chronology of deposition: during LTC1/C2 (c250-175BC) deposits are rich in weaponry, human and animal remains, followed by a hiatus until LTD2a/D2b (c70-50BC) when deposits resume but are less military in character, consisting mostly of coins, jewellery and pottery. A few sites seem to date from after the conquest, between 50 and 20BC, and these appear to be already at least partially Romanised, using masonry architecture.

Any evidence of settlement abandonment also dates up to fifty years before Caesar. Farmsteads were being abandoned in the Aisne Valley in the later second and the first half of the first century BC (Haselgrove, 1996a) and Bayard (1996) notes the same pattern in Picardy between the start and middle of the first century BC. Of the 50 or so sites surveyed, about a quarter of the total seem to have been abandoned between the end of LTC and about 50BC. Three quarters of the sites survived the Gallic War, but few of these survived for longer than another hundred years. Bayard's figures for abandonment are generally lower than those of Haselgrove & Scull (1992), but take into account settlements which have shifted slightly during the period.

> *"On remarquera que les sites de la fin de la protohistoire sont caractérisés dans cette petite région par une grande instabilité. Cette instabilité se traduit parfois par des désertions, parfois par un déplacement de quelques dizaines à quelques centaines de mètres"*
>
> (Bayard, 1996:166).

Sites which do appear, with the exception of the fortified elite centres mentioned earlier, do not display any particular defensive characteristics. Many are enclosed by ditches, but these seem to be more for drainage and demarcation than for defence, being generally no more than a couple of metres in width, with relatively low, unrevetted banks. Once again, the impetus towards such enclosures predates the Caesarian period, beginning in the late second and early first century BC. There may be a trend towards agglomeration, for example at Beaurieux where a large enclosure appears to have housed several separate farm compounds (Haselgrove, 1990b). Haselgrove attributes such larger groupings to a crisis response.

> *"The process of nucleation during late La Tene may have occurred on more than one level, with a number of single farmsteads gathering together for protection around the dwelling of the local leader due to the generally unstable conditions, quite independently of the construction of larger fortified centres during particular emergencies. Alternatively, these developments may have followed the return of the population to the countryside in the post-crisis period, recent events leading them to rebuild their former settlements as small enclosed hamlets rather than single farms" (ibid:258).*

Any changes in settlement location are explicable in similar terms. Despite the instability of settlements, the picture of the subsistence economy seems to be much more constant and enduring. There is some evidence of agricultural intensification from a decline in tree pollen during the second and first centuries BC (Defgnée & Munaut, 1997). Minor changes noted include a preference for pig meat at the oppida, but no overall changes are evident (Lepetz, 1996). Changes in trade patterns have been discussed above. No unoccupied zones appeared.

It seems, therefore, that very little preparatory action occurred before the Gallic War - hardly surprising, given the speed of events. Preparations for defence made fifty or more years earlier seem to have been re-utilised and in one or two cases, improved or repaired, at the last moment, but it seems clear that the people of the region were generally unprepared for the size and nature of the Roman attacks, which were of an order and type never before experienced in the region.

Presence of functional correlates

Perhaps more surprisingly, the functional evidence for warfare during the 50s BC in Gallia Belgica is as elusive as any causal or preparatory evidence. Despite the vast numbers of combatants mentioned by Caesar, and his accounts of punitive actions, including the burning and destruction of villages, crops and stores, no such evidence has yet come to light that can be unequivocably assigned to conflict. At Riemst-Lafelt, in Limburg, pits have been excavated containing burnt pottery, and one large pit was completely filled by burned wattle and daub, at a site subsequently abandoned and then later occupied by a Gallo-Roman villa, but the heavy erosion of the surfaces at this location preclude any certainty of aggressive destruction (Bourgeois, J. et al, 2003). There are no examples of unburied bodies or mass burials, no examples of weapon trauma, no damaged weaponry. This lack has been noted by many commentators (e.g. Hamilton, 1996; Haselgrove, 1990a and Wightman, 1985). It is perhaps not unexpected that little functional evidence is recovered from the Aisne Valley, the focus of much of Haselgrove's research.The Aisne was the territory of the Remi, who very quickly allied themselves with Caesar against their neighbours, and therefore presumably saw little in the way of military action or destruction. However, events such as the siege of the Atuatuci stronghold (either Namur or Mont Falhize near Huy, on the Meuse) should be archaeologically traceable.

A few sites may represent Roman camps utilised during the prosecution of the campaign but remain without secure dating, and as Haselgrove has pointed out (1990a, 1990b) Caesar probably used existing native sites for garrisons whenever possible.

Explanations for the lack of evidence for a war that apparently cost so many lives and saw so much destruction are at best tentative and are generally not attempted in the literature. Perhaps the sheer speed, first of the campaign and then of the Romanised takeover of land and economy, precluded the formation of archaeological stratigraphy. We can possibly imagine mass pyres of bodies, rapid clearance and rebuilding of farms, and destruction or removal of weaponry as the Roman administration operated to begin creating profit from the newly conquered territory. Efficiency on such a scale, capable of removing all the evidence, seems incredible, however, and is not borne out by similar evidence from other Roman provinces, where at least some trace of conquest has been recovered. In this respect, Gallia Belgica remains a mystery.

Presence of resultative correlates

Changes subsequent to the Gallic War are much more evident in the archaeological record. In these cases, the problems are less of finding evidence, more of being able to tie the evidence to the occurrence of warfare.

Changes in subsistence patterns are traceable in the decades following the conquest in some areas. There is an apparent drop in tree pollen - from 16% of average totals before 50BC to 12% between 50BC and the Flavian period. Tree pollen that occurs includes a large amount of *Corylus* and *Betula*, especially at indigenous farms, which are pioneer genera which rapidly colonise rough land, abandoned meadows and pasture, and boundaries (Defgnée & Munaut, 1996). Gallo-Roman sites have relatively little concentrations of these genera, perhaps indicating greater clearance and land utilisation, and clearance of scrub on those estates. Similar increases in these two genera and *Carpinus* are noted by Kreuz (1999), along with a decline in *Quercus* and the first appearance of *Juglans regia* and *Castanea sativa.* It is claimed that there was a part-regeneration of oak and pine forest, and decolonisation of scrubland, in the first decades after the conquest until the beginning of the first century AD (Carroll, 2001). However, the time-scales for changes are hard to determine with any certainty within a period of several decades. Kreuz (1999) also notes the decline in pasture (particularly wood pasture) weeds such as *Plantago,* Chenopodiaceae and *Pteridiam aquiline* but an increase in weeds of hay meadows including *Papaver, Artemisia sp.,* and *Rumex acetosella.* Cereal pollen appears to increase immediately following the conquest, possibly related to a policy of grain production for army use (Kreuze, 1999). Rye (*Secale* cereale) only appears after the Conquest as a regular crop; concentrations of other cereal pollens show a shift in preference. At Iron Age sites, barley (*Hordeum sp.*) was the predominant grain, followed by lesser quantities of *Triticum dicoccum* (emmer) and *Triticum spelta* (spelt). By the Roman period, the quantities of barley are still the largest, but not as high, and spelt has replaced emmer as the second most popular cereal. *Triticum aestium* and *Triticum durum* make their appearance at the same time (Kreuz, 1999). It is suggested that spelt produced higher yields and was therefore the grain Romans preferred with a view to feeding the army. Spelt is a winter wheat, and its preference may explain the increased presence of winter crop weeds such as *Agrostemma githago*, which becomes common on most Romanised sites. Grain storage pits become increasingly rare during LTD2 and in some areas at least are absent by the time of Augustus and Tiberius (Lambot, 1996). The storage of grain in large above-ground *horrea* rather than in small 4-post granaries or pits may also explain the increase in *Sitophilus granaries,* the 'granary weevil', in assemblages (Kreuz, 1999).

During the Iron Age, there is little evidence for the growing of spices, vegetables and fruits in gardens, but this changes after the conquest, with the growing of imported species such as *Prunus persica* (peach), *Prunus domestica ssp. institia* (plum), *Coriandrum sativum* (coriander) and *Apium graveolens* (celeriac) among others. Imported exotic species also begin to occur on domestic sites after the conquest - *Phoenix dactylifera* (date), *Olia europaea* (olive), *Pinus pinea* (stone-pine) and *Piper nigrum* (pepper) (Kreuz, 1999). Kreuz feels that there is "no doubt that the differences in the plant species cultivated and used result from a change of diet from Iron Age to Roman times" (1999:85).

The pattern of stock also appears to change, not least in the size of animal farmed. There is an evident increase in the size of cattle of some 20% (from an average pre-Roman wither height of 1.10m to up to 1.43m by the first century AD). Study of the morphology of these animals suggests an origin from outside Northern France. Gaulish animals have a foot base that is proportionately large in relation to the limbs, while the larger animals have more developed limb bones in relation to foot size. They are more robust and tend to have more developed musculature (Lepetz, 1996). Pigs, sheep, chickens and horses are also seen to be larger and more robust, and there is a recorded increase in the forms of domestic dogs. Haselgrove (2007:503)

	Germanic period	Roman period	Modern
Cattle average height	109cm	124cm	135cm
Horse average height	130cm	140cm	160cm
Pig average height	63cm	63-82cm	90cm
Sheep average height	54cm	60-71cm	75cm

Table 11: Comparison between average stock heights for the Germanic and Roman periods (after Kreutz, 1999).

speculates that the appearance of larger horses, introduced probably from the south of Gaul, at Condé-sur-Suippe and Montmartin, was due to Roman influences in the decades before Julius Caesar's campaign. These changes may have just started to occur at the end of the La Tène period, but they become much more evident and widespread after 50BC. The appearance of Romanised urban centres and the military market may have stimulated breeding changes. There are also indications of an increase in the use of cattle for traction, such as an increase in the number of animals slaughtered at an age of seven years or more, notably on *villae* sites (ibid).

Further research is needed to establish the details and concentrations of agricultural change between the late Iron Age and the Roman periods.

> *"A change of dietary customs might or might not imply a change of group or ethnic identity. This is very difficult to prove. But one thing is certain: if a group changes its dietary customs there must first be a change in production and/or the mechanisms for the provision of food. If such changes were brought into a region by new inhabitants, it would have economic and ecological consequences for the landscapes under investigation, and that is what we can look for"*
>
> (Kreuz, 1999:80).

There is also an instance of cessation of production of another commodity - salt. At Pont-Rémy, about 10km from the sea on the Somme estuary, a large salt producing site which had developed from LTC1, gaining enclosure banks, ditches, houses and a small attached cemetery, and by 100BC had the possible capability of producing between 6-12 tonnes of salt in a six month season, declined and was abandoned by LTD2. In the Flavian period a villa was built 800m to the west of the site (Prilaux, 2000). Many rural sites seem to have undergone a major transformation after the conquest, but for most, these changes did not occur until the earlier half of the first century AD. This topic will be examined more fully below.

The establishment of the Rhine frontier, with its permanent military bases and subsequent urban centres, in addition to the urbanisation of native centres such as Reims and Soissons, had an effect on the pattern of trade and markets in the region. An important element was the construction of major roads connecting Romanised central Gaul with the Channel ports and the Rhine. The Augustan period saw the first major phase of road construction, with sections known from excavation and survey including a link from the road southwards to Lyon from Trier via Metz and Toul, a road heading towards the Channel that linked Senlis, Beauvais and Amiens, another between Reims, St. Quentin, Arras and Therouanne, and a spur heading east-west between this road and the towns of Bavay and Tongres. Another road was under construction along the west bank of the Rhine, connecting Köln, Xanten, Nijmegen and Voorburg. It is along these routes and in the settlements founded along them that most evidence for Roman goods and coins tends to be found (Haselgrove, 1987). It has been suggested that, initially at least, control of the network of trade was left in the hands of ruling native elites (those who had made accommodations with the new Roman administration), a picture borne out by the concentration of Roman luxury goods in late first century BC aristocratic graves and by the continuing minting of certain Gaulish coinages.

> *"It must be assumed that the duration of this network in purely native hands was short, and was terminated by the commencement of the Roman campaigns against Free Germany in the penultimate decade of the first century BC. It seems likely that Rome took this opportunity to tighten control over Belgic Gaul ... and that Strabo's comment that its civitates were 'now for the most part subject to the commands of Rome' (Geographica IV, 4.3) is a thinly veiled 'apology' for its final annexation"*
>
> (Haselgrove, 1987: 119-120.

Allowing the aristocracy to maintain control of exchange, particularly with those populations still outside Roman control (in Germany and Britain) permitted the Romans to leave status and therefore a measure of social control to the tribal leaders in the short term, and also provided a mechanism for the acquisition of goods from unconquered territories that otherwise may not have been available. The wealth accrued by the elites (particularly, for example, by the aristocracy of the quisling Remi) allowed them a further measure of control over the swarm of Roman merchants, traders and money-lenders that arrived soon afterwards. In the more northerly regions that had suffered to a greater degree from the depredations caused by the conquest, the evidence for trade is lacking, and we can possibly assume that here, the pattern of trade was mostly of export (perhaps by levy) of agricultural produce (and perhaps salt) without the balancing benefits of imports and wealth accumulation.

There is no evidence for any trophy collection or war memorial building in the region, nor is there any surviving post-war iconography, except for the occasional military tombstone. Literary and epigraphic evidence for the presence of Roman military forces has been summarised by Wightman (1974). However, Roman coins in the decades after the Conquest utilise depictions of Gallic prisoners and Gallic weapon trophies as a form of propaganda (Wells, 2007).

The demographic evidence for the post-conquest period is very scanty. There is no information about the sex and/or age ratios of the surviving populations and little to suggest the depopulation that Caesar's own figures would suggest. Certain areas of the region show much slower recovery and lower status than others, and it is possible that it is in these places that a reduced population struggled to exist at a reduced level of subsistence - Haselgrove has suggested that this could be the case in the lands of the Nervii (1990a:58) although much depends on a better understanding of the size and boundaries of tribal territories. The problem is compounded by the general invisibility of burial practice in the period, except for a very few high status examples.

Unoccupied zones are not evident; there may, however, be some evidence of the transplantation of populations, perhaps into areas that had suffered massive depopulation as a result of the reprisals visited upon certain tribes by Caesar's troops. It seems that at least two tribes, the Eburones and the Atuatuci, disappear entirely from the historical record after the conquest, and they seem to be replaced by the Batavi and the Tungri respectively, people first mentioned by Pliny, and referred to as the 'sworn ones' (Wightman, 1985). The Eburones were the subject of vicious attacks in 53BC, their villages destroyed by fire and their lands plundered. Their cattle were driven off and their crops destroyed. A number of settlements on the Germanic side of the Rhine associated with this tribe appear to have become permanently abandoned at this time, including Niederzier, Eschweiler-Laurenzberg and Euskirchen-Kreuzweingarten (Carroll, 2001). The region seems to have been reoccupied in the Augustan period; it is suggested that a dangerous gap had been left by the emptying of this area, making it prone to attack and occupation by unfriendly Germanic migrants, and presenting a threat to the Roman pacification and reorganization of the region. For this reason, according to Tacitus and others, the friendly Germanic Ubii from the Lahn valley east of the Rhine were transplanted into the old Eburones territory. Subsequently, the Batavi settled north of the Tungri and Ubii in the eastern Dutch Rhine delta. Evidence for the appearance of the Batavi may include the appearance of new building styles with beam-slot walls, as found, for example, at Oss-Ussen in North Brabant (Schinkel, 2005) and a number of silver coins - one group from eastern or central Gaul inscribed TOGIRIX (the only examples of this type known in the north), and a group of silver *triquetrum* coins of the Batavi found at Empel (North Brabant), where a later Romano-Gallic temple was founded, dedicated to Hercules Magusanus, the chief god of the Batavi (van den Broeke, 2005). The Ubii, prior to the Roman invasion, are said by Caesar to have been under threat from the much stronger Suebi and had been forced to pay tribute to them, so they were happy to throw their lot in with the Romans. A new settlement, Ubiorum, was founded on the Rhine for a group of this tribe, which was later promoted to the level of colony - the Colonia Claudia Ara Agrippinesium (Köln) - in AD50. At a number of other sites on the west bank of the Rhine, there is no pre-Roman trace of settlement, and the communities all seem to have been founded around 30BC. They include Neuss, Bonn, Dormagen and Remagen (which has a timber palisade dendrochronologically dated to 9BC), and may represent a series of assisted colonies of the Ubii supported by Roman policy (Carroll, 2001).

In Belgium, a number of excavated Late Iron Age sites appear to go out of use around the middle of the first century BC, while some show evidence of continuation into the Roman era, and others are replaced by villas or *vici*, or are founded *de novo* around the period of Augustus. The oppidum site of Wallendorf was abandoned around 50BC but was later developed into a *vicus* and Gallo-Roman sanctuary (Krausse, 2003).

The following table brings together a brief overview of the evidence for settlement occupation and abandonment in Belgium from the Late Iron Age into the Roman period. There is no clear pattern to the sites - those which survived or were replaced in the Roman period tend to be on slightly better soils or are nearer to communication routes, but not all cases are clear-cut in this respect.

The region comprising Picardy, Artois, parts of the Champagne and Seine Maritime in Normandy, studied by Haselgrove, showed a general pattern of settlement increase until La Tène D2 (90/85-30/20BC) when abandoment became relatively widespread (Haselgrove 2007:503). Whilst pollen evidence mentioned earlier has pointed to change of use of tracts of countryside, there is very little evidence for any abandonment of developed arable land. Indeed, it would have been in the interests of the Romans to maximise grain production as quickly as possible to supply their troops and fortresses. However, reorganization may have led to some small relinquishment of parts of former farming territories. At the Actiparc site six kilometres north-east of Arras, a new Roman road divided an older group of farming settlements, and the western zone became uninhabited and under the control of the military unit, but it is unclear whether or not it was still being exploited for arable (Blancquaert & Priluax, 2003).

The post-Conquest period saw numerous changes in settlement location and morphology. The imposition of the

Figure 18: Settlements and other sites in Gallia Belgica mentioned in the text

1. *Kemmelberg*
2. *Breteuil*
3. *La Verberie*
4. *Champlieu*
5. *Condé-sur-Aisne*
6. *Mauchamps*
7. *Somme-Bionne*
8. *Eprave (Trou de l'Ambre)*
9. *Court-St. Etienne*
10. *Mauberge*
11. *Tongres*
12. *Haps*
13. *Pommiers*
14. *Erden*
15. *Niderburg*
16. *Sint-Anfries*
17. *Osstkamp-t'Swarte Gat*
18. *Oedelem-Wulfsberge*
19. *St. Martens-Laten-Brakel*
20. *Deinze-RWZI*
21. *Aalter-Langevoorde*
22. *Zele-Zuidelijke Omleiding*
23. *Berlare-Kerkveld*
24. *Lede-Leedshouwken*
25. *Petit Enghien*
26. *Brugelette-Bois de l'Attre*
27. *Laudeuze*
28. *Riemst-Lafelt*
29. *Tongeren-Henis*
30. *Stevoort*
31. *Donk*
32. *Tienen-Wijngaardenberg*
33. *Hoegaarden-Goudberg*
34. *Orp-Le Tierceau*
35. *Hannut-Trommelveld*
36. *Gingelen_Melenveld*
37. *Kontich-Kapelle-veld*
38. *Kontich-Altsberg*
39. *Wijnegem-Steenakker*
40. *Ekeren-Het Laar*
41. *Brecht-Overbroek*
42. *Meer-Zwalumstraat*
43. *Meet-Hoog Eind*
44. *Ribemont-sur-Ancre*
45. *Braives*
46. *Wederath*
47. *Velzecke*
48. *Courtrai*
49. *Tournai*
50. *Empel*
51. *Oss-Ussen*

road system and urbanised population centres was crucial to the continuing control of the region. "Practically all the civitas capitals of north Gaul were established, without regard to pre-existing centres of Gaulish population, along these routes, which had been laid out primarily for military purposes. Examples of this are Atuatuca-Tongres (*civitas Tungororum)*, Bagacum-Bavay *(civitas Nerviorum)*, Castellum-Cassel *(civitas Menapiorum)*, and Tervanna-Therouanne *(civitas Morinorum)*" (Mertens, 1983:155). These centres were often a few kilometres away from the previous Iron Age settlements - Arras, Reims and Soissons - and sometimes a larger distance - St. Quentin and Avenches. The preference for the siting of these new settlements was at bridges, crossing points or road junctions. Many sites were abandoned, particularly smaller fortified sites while

Site name (Province)	Date	Location	
Aalter-Langevoorde (EF)	LIA-C3AD	clayland	traces of LIA settlement; C1AD large enclosed settlement established (ferme indigene)
Berlare-Kerkveld (EF)	LIA	sand ridge	LIA settlement traces, abandoned?
Bierbeek-Grote Heide (FB)	LIA	loamy sand	LIA settlement traces, abandoned?
Brecht-Overbroek (A)	LIA	sand ridge	LIA settlement abandoned; 500m to east is important Roman vicus settlement
Beugelette- Bois de l'Attre (H)	multiperiod to LIA	plateau site	large multi-period settlement; no traces of settlement in Roman period
Chievres (H)	LIA to early Roman	plateau	LIA settlement with occupation into early C1AD?
Deinze-RWZI (EF)	mid-final LIA, early Roman	sandy ridge	LIA building; traces of Roman agricultural and funerary activity
Donk (L)	LIA	loamy sand ridge	settlement ends mid C1BC; Roman occupation C1AD-C5AD
Ekeren-Het Laar (A)	LIA (C1BC)	sand dune	large settlement; abandoned? Later colonisation by Batavi possible (house type change)
Gingelom-Miolenveld (L)	LIA	valley slope	ferme indigene?
Hannut-Trammelveld (Le)	LIA	hilltop loess	occupation and enclosure; abandoned?
Hoegaarden-Goudberg (FB)	LIA	loam/loess slope	eroded settlement; subsequent Roman villa
Kontich-Alfsberg (A)	LIA	plateau	occupation from La Tene 1; late phase large earthwork fortification, destroyed in early Roman period; abandoned. 400m from vicus at Steenakker
Kontich-Kapelleveld (A)	LIA	plateau	occupation traces; abandoned
Ladeuze (H)	LIA, C1AD	plateau	quadrilateral palisaded enclosure, settlement; occupation ends late C1BC, enclosure destroyed. Two C1AD Gallo-Roman graves
Oedelem-Wulfsberge (WF)	LIA - early Roman	clay/sand ridge slope	LIA and early Roman settlement; ferme indigene?
Orp-Le Tierceau (WB)	LIA	loess ridge	LIA occupation; abandoned?
Petit-Enghien (H)	LIA to early Roman	loamy plateau	LIA occupation; burned daub and outbuilding; possible palisade, early Roman period pottery
Riemst-Lafelt (L)	LIA	loess plateau	much eroded site with traces of LIA occupation, burned wattle-and-daub, burned pottery; site later occupied by Roman villa
Riemst-Valmeer (L)	LIA to early Roman?	loess plateau	traces of LIA activity; later villa
Sint-Andries (WF)	LIA	sand ridge	ritual site? no reoccupation
Sint-Martens-Laten-Brakel (EF)	early Roman	sand dune	Gallo-Roman settlement, no LIA evidence
Stevoort (L)	LIA	sand/loam river bank	settlement; abandoned?
Tienen-Wijngaardenberg (FB)	LIA	loam ridge	settlement; abandoned?
Wijnegem-Steenakker (A)	LIA to early Roman	sand ridge	settlement founded after 50BC? Vicus
Zele-Zuidelijke Omleiding (EF)	LIA to early Roman	sand ridge	settlement continued into Roman period

A: Antwerp *EF: East Flanders* *FB: Flemish Brabant* *H: Hainaut*
Le: Liege *L: Limburg* *WB: Walloon Brabant* *WF: West Flanders*

Table 12: Late Iron Age - Early Roman settlement sites in Belgium (data derived from Bourgeois, J. et al, 2005:191-299; and Namur: Division du Patrimoine, 7/1999)

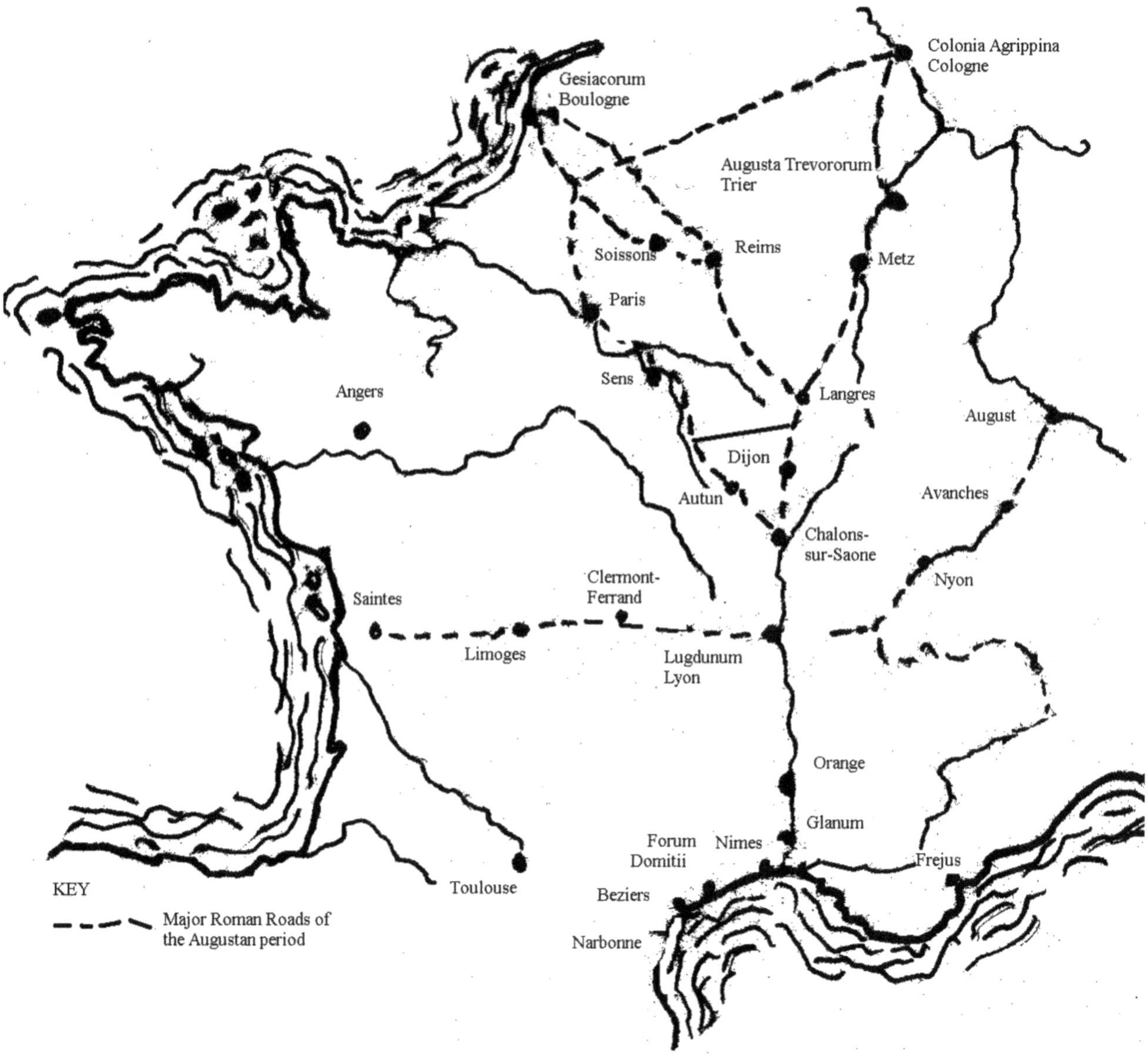

Figure 19: Roman roads in Gaul (after: Blagg & Millett, 2002)

some native oppida survived the conquest, but few lasted for long. A few became small Gallo-Roman *vici.*

> *"The disappearance of fortified settlements was probably connected to the introduction of a new system of administrative control and the stationing of Roman troops in the Rhineland by Augustus. This enabled the Roman government to stop the system of intertribal warfare, thus rendering obsolete heavily fortified settlements located on difficult-to-reach hilltops"*
>
> (Roymans, 1990:211).

Roymans goes on to suggest that this was part of a process of detribalisation, which "ran parallel with the process of integration into the Roman Empire" (ibid:268) which only really began in the Augustan period, after the Batavian Revolt. In place of warfare as a means of socio-political competition, the Roman system introduced the market economy and participation in commercial and administrative activities. By this means, the tribal warrior elites were 'socialised' into the Roman state system, and the rural populations, who had benefited from "tribal forms of exchange, reciprocity and redistribution" were turned into peasants, a process "represented archaeologically by the development of the villa system in the countryside" (Roymans, 1990:269).

The major periods of building at the new centres were during the Augustan period, and then a resurgence in the Flavian period. At some sites, evidence of regular street grids has been excavated (at Reims, for example), and forum-basilica complexes (Haselgrove, 1990a). V-shaped ditches have been traced at a number of locations, possibly indicating a military presence at the time of their initial

	Evidence of pre-Roman occupation	Evidence for new Roman foundation	Connection with Roman roads	Military-style ditches
Amiens	no	no (until later)	yes	no
Arlaines	no	yes	yes	yes
Arras	no	yes	yes	yes
Autun	no	yes	yes	no
Bavai	no	yes	yes	no
Braives	no	yes	yes	yes
Elewijt	no	yes	yes	yes
Liercourt-Hérondelle	yes	no	no	yes
Mt. Beuvray	yes	no	no	no
Pommiers	yes	no	yes?	no
Reims	slight	yes	yes	no
St. Quentin	no	yes	yes	no
Soissons	no	yes	yes	no
Therouanne	no	yes	yes	no
Titelberg	yes	no	yes	yes
Tongres	no	yes	yes	yes
Tournai	no	yes	yes	yes
Velzeke	no	yes	yes	yes

Table 13: Selected sites in Early Roman Gaul and their relationship to pre-Roman settlements and roads (after Wightman, 1977, with additional data included)

construction. Tongres has been dated to the 20sBC and was enclosed by such ditches. Tournai is probably contemporary, as are Elewijt, Braives and Velzeke, all enclosed by similar ditches (Mertens, 1983). These sites have little or no evidence for any earlier occupation. It is possible that in the early phases of a number of these sites, control was in the hands of Roman military personnel. There are possible Caesarian forts at Liercourt-Érondelle, Berry-au-Bac and Mauchamp (Haselgrove, 2007:512) and Roman troops were probably stationed at La Chaussée-Tirancourt and Titelberg. The evidence suggests that the Roman camp at La Chaussée-Tirancourt was built by Gaulish auxiliaries between 40 and 20BC on a site which may have been occupied earlier. The oppidum at Pommiers seems to have been founded around 50BC, possibly succeeding the site at Villeneuve-Saint-Germain five kilometres away, but in turn, this site was abandoned in the Augustan period in favour of Augusta Suessionum (Soissons) (ibid, 510).

Settlement location changes outside the urbanised centres are also visible, but here there is more ambiguity. As has been previously noted, the first century BC has been characterised as a period of settlement instability, and it is often hard to be certain that abandonments or new foundations can be directly related to the conquest. However, Walthew (1982) regards the appearance of large numbers of Romanised villas in the region of the Somme in the early years of the first century AD as "part of a concerted and systematic Roman attempt to encourage, if not actually to impose, massive rural development in at least one part of Gallia Belgica from an early date" (ibid:232). This kind of rural investment is lacking in less fertile parts of the region - on the sands of Flanders west of the River Scheldt, there are very few villa establishments, or indeed towns. Here, a change of settlement pattern is observed from the earlier decades of the first century AD in the form of a number of small, relatively poor, settlements concentrated along roads and navigable waterways, such as Courtrai, Harelbeke, Waasmunster and Kruishouten (Vermeulen & Bourgeois, 2000). None of these sites has earlier occupation strata.

In the valley of the Aisne, there may have been an increase in the number of foundations of rural sites marking a gradual return to the countryside in the 'improving conditions' following the Roman conquest, with at least nine new sites of the late first century BC being identified in the survey area, in contrast to the pattern of abandonment of sites occupied before the arrival of Caesar (Haselgrove, 1996a).

The recent discoveries at the Actiparc site *(figure 20)* between Lille and Arras demonstrate another form of settlement change - the militarised control zone. This site acquired a Roman military post, possibly sited to control the upper valley of the River Scarpe in Atrebates territory. Here, the excavators consider that the construction of the military fort and its access road represent the beginning of a complete reconstructing of the countryside under the control of military authorities. Beginning with wooden barracks and other structures enclosed by V-shaped ditches, in a trapezoidal plan, with an access road lined with palisades, the fort was soon strengthened with palisaded enclosures, exterior defences and angle-towers. A short way down the road, a further palisaded enclosure surrounded a large number of granaries and a hamlet of Gauls. This enclosure was one section of a larger emplacement at least three hectares in extent, which included a very large open area and at least three other enclosed areas. A road ran north to south across the complex, lined by palisades " destinée à faire converger less passants vers des points de passage obligé" (Blanquaert, Prilaux et al, 2003:23). "Cet ensemble où présence militaire et activité écommuniqué sont étroîtment imbriquées est donc un étonnat témoignage des directives données par les autorités romaines pour pacifier la région, mais aussi favoriser son développement économique" (ibid). It is possible that other such 'militarised zones' await discovery.

Morphological changes in settlements include the regular street grids found in the civitas capitals (along with fora and basilica), and a pattern of more regular rectilinear enclosure and building orientation in some post-Conquest rural settlements, although native architectural styles certainly persisted until well into the first, and sometimes into the second century AD (Lambot, 1996). In the northern lowlands, settlements lost their previous instability and adopted more regular layouts, mirroring a pattern found before the Conquest in the more southerly part of Gallia Belgica (Roymans, 1990). A group of larger enclosed settlements, comprising a number of separate dwellings with their own yards, pits and wells, laid out round a large open space (e.g. Bazoche, La Foulerie II, Beaurieux Les Grèves (Aisne) and Vermeuil-en-Hallatte) "appear to be a post-conquest development" (Haselgrove, 2007:506). Beaurieux seems to have been abandoned for one or two generations in the mid first century BC and then reoccupied during the Augustan period (ibid:508).

Changes discernable as a result of the movement of refugees are not found, nor are any accounts of refugee movements other than the retreat of the Morini and Menapii into forests and marshes in the north of the region, although there is a possible suggestion that some of the people of Gallia Belgica made their way either to the east of the Rhine or across the Channel. Biodiversity changes as a result of other forms of relocation have not yet been identified, although Caesar records raids by Germanic tribes into the lands of the Menapii in 55BC, which presumably had effects on local environments.

It is certain that for at least a generation after the conquest, sections of the surviving population would have undergone great hardship. In 54BC Caesar mentions a drought and poor harvests in Gaul, and he records the burning of houses, stores and crops, and the driving off of stock, on several occasions between 56 and 53BC. Given the claims for the number of dead and people sold off into slavery (53,000 of the Atuatuci who had been besieged at or near Namur were sold in one lot), it is only reasonable to suppose that there was a shortfall of workers to provide food, and much suffering caused by homelessness and starvation. However, there is no corroborating evidence for this, other than the rather slow recovery and low status of settlements in the worst affected regions.

The appearance of Roman imports, both artefactual and cultural, is readily discerned, although once again it should be remembered that this process had already begun via trade in the period before the conquest. There are many mechanisms of social change which can be reflected in artefactual and cultural change; identifying those directly resulting from aggression is not simple. Much of the visible change in Gallia Belgica could be the result of other processes - acculturation, non-violent assimilation, changes in access and control and so on (Okun, 1989). In the Netherlands, as elsewhere, the "gradual replacement of a large part, or even all, of the local, domestically produced pottery by Roman wares is ... one of the clearest signs of the degree of specialisation increased in this part of the country after the Iron Age" (Van den Broeke, 2005:690).

At the Titelberg, a large oppidum, occupation seems to have declined around 20BC, but at the turn of the millennium, roads are paved, Roman-style materials are used to rebuild houses, and whilst local pottery continues, Roman-style wares become more numerous (Hamilton, 1996). Roman military occupation has been suggested at a number of civitas centres by the presence of pottery and fibula types, as well as some military material, but the arrival of any Roman goods in rural sites was very slow, and in the southern Netherlands, the Caesarian campaigns seem to have had little visible impact, the major changes discernable being attributed to the later campaigns of Drusus around 12BC (Van den Broeke, 2005).

A pottery industry appears to have been developed *de novo* along the Rhine in the later C1BC or early C1AD, apparently the result of external initiatives, with no relationship to indigenous forms (Drinkwater & Vertet, 1992). Elsewhere, apart from in the civitas capitals, local forms of pottery continued to predominate well into the first century AD.

Much discussion of the coinage of Gallia Belgica has been undertaken. The region had begun to produce local coinage by the end of the second century BC, with potin and precious metal issues becoming widespread in the first century BC. During the decades between Caesar's invasion and about 15BC, numerous hoards of coins were deposited, and these demonstrate a decline in weight and precious metal content, ("In the course of the war, both the weight and standard of gold coinage fell dramatically; gold experienced a debasement from a purity of over 60 per cent to 45 per cent or even less" (Haselgrove, 1984:113)),

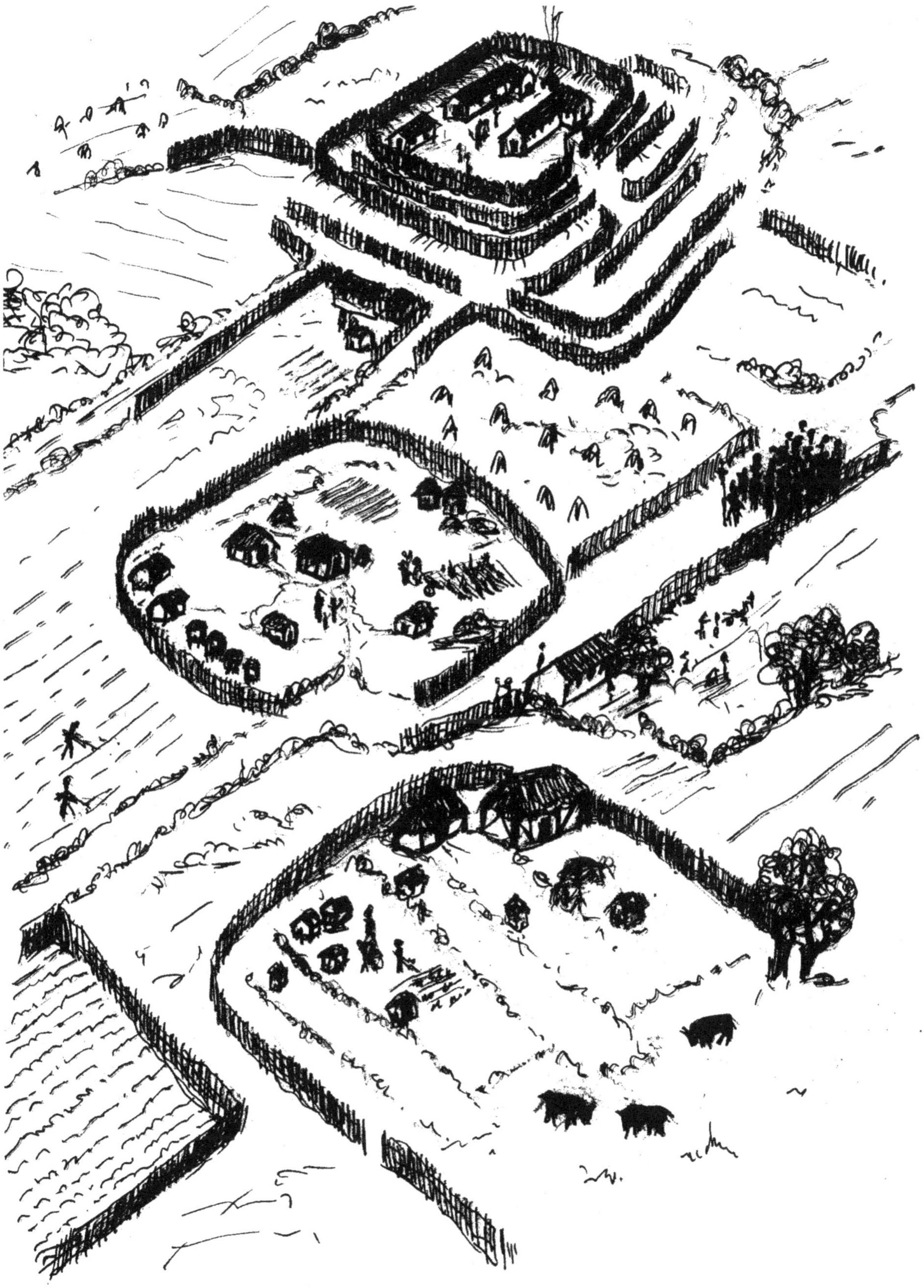

Figure 20: Hypothetical reconstruction of the Actiparc site near Arras (after Archeologia No.404, Octobre 2003)

which can be taken as a suggestion of a "time of troubles" (Wightman, 1985:38). Central Gaulish silver coinages spread into Gallia Belgica during the war years, possibly because the Romans used these issues to pay Gaulish auxiliaries accompanying the legions. "There is a rough correspondence between areas of concentration and the legions' winter quarters" (ibid:39).

After the Conquest, Roman coins remained relatively rare in the region, and native monetary systems were apparently allowed to develop for some decades. Precious metal issues declined, although the Remi were still striking gold coins, and a silver type of the Treveri is known. Bronze issues became common for the first time. This may be a result of a shortage of precious metals caused by Roman seizure for booty or tribute, or as a result of a drain of precious metal coinage in previous years in order to pay for mercenary assistance during the war (Wigg, 1999). In addition, much coinage was deposited in unrecovered hoards, and some may have been taken as personal wealth by leaders who fled the region (Haselgrove & Scull, 1992). The influence of Rome can, however, be seen in certain typological aspects of the coins, such as the creation of copies of denarii showing elephants. After 15BC there was a sudden massive influx of Roman bronze coinage, and for a while both Roman and native coins were in circulation, although Roman military and urban sites seem to have tended to retain more Roman specie than the rural sites.

By allowing the continued production of local coinages (which are present in a very large range of localised types), the Roman authorities may have been able to achieve a number of useful goals. Auxiliary and mercenary troops could be paid without causing a drain on official Roman monetary sources, local elites could be rewarded through allowing them to be able to make and control money, the concentration of local wealth in the hands of these elites would allow them more social control, thus saving the Romans the need to police areas, and the coinage provided a mechanism for the payment of tax and tribute by the populations of the conquered regions (Wightman, 1977; Haselgrove & Scull, 1992).

Further indications of artefactual and social change can be discerned in forms of ritual and burial practice. Continuity of cemetery use in rural sites has been noted for Flanders (Vermeulen & Bourgeois, 2000), and in Le Porcien (Lambot, 1996) where graves of immediate post-conquest decades are notable for their poverty compared to previous times.

Haselgrove (1984:115) has noted a marked differentiation in mortuary rite which develops in restricted areas of northeast Gaul at this period against a background of cremation burial, particularly along the Aisne and Mosel which have a juxtaposition of Roman imports with traditional symbols of wealth and authority, a practice with pre-Conquest beginnings but most undoubtedly post-Conquest date. Few other graves have been found in this region. The elite grave ritual may have been a way of projecting the desirability of alliance with Romans and Roman practices for the ruling class. "It is possible that the ruling groups of the allied and free states were, in fact, allowed to maintain an armed retinue, and weapons were certainly interred in the burials of the elite down to the end of the first century BC" (ibid:113).

At the large cemetery of Wederath-Belginum, 40km west of Trier, a decline in the amount of female jewellery in graves has been noted during the later part of LTD2, and around 20BC Roman coins and Mediterranean artefacts begin to appear as grave goods. At about the same time, the number of enclosed grave "gardens", possibly family plots, increases. Despite a population increase during the first half of the first century AD, perhaps partly composed of immigrant traders or officials, and a scattering of Roman or imitation-Roman artefacts, the majority of graves appear Iron Age in form and content (Hamilton, 1995). Wells (1999) has noted the incidence of "un-Roman" burial rituals at, for example, Goeblingen-Nospelt in Luxembourg, and has suggested that a refusal to adopt the new form of rite may be seen as a form of resistance to the Roman occupation.

Wellington (2001) has identified three chronological groups of temples or sanctuaries in Northern France. The first group dates to LTC and few survive into later periods; the depositions consist of much weaponry along with human and animal remains. The second group begins in LTD (c. 70-50 BC) and in these examples, pre-Roman structures are found, sometimes under later Roman stone buildings. Depositions at this group are more varied, including coins, pottery, and brooches as well as bones, but fewer weapons. The third group dates to the early Roman period, between 50 and 20 BC, and this group lacks any pre-Conquest deposits apart from coins. They generally have stone structures, with no trace of any preceding wooden buildings. Large temples are a rarity in the urban centres of Northern France in the early Roman period, but monumental building at rural sanctuaries appears to have begun before the middle of the first century AD.

Six sanctuary sites have been excavated in the Middle Rhine and Moselle region. Of these, Wallendorf has little late LT evidence but a new sanctuary was built in the Augustan or Tiberian periods, and new sanctuaries were built around the same time at the Martberg (dedicated to Lenus Mars), Dalheim and Bastendorf. Gallo-Roman sanctuaries were built on what may have been Iron Age ritual sites at Belginum and the Titelberg, where finds include brooches, miniature weapons, thousands of animal bones and fragments of human skulls.

> *"It seems that the construction of temples in the region of the Treveri did not start before the cultural change which came about as the result of*

Roman influence in the post-conquest period. This is especially clear in the case of the huge fortified settlements ... where the main floruit of the temple complexes obviously coincided with the progressive loss of the political and economic importance of these settlements"

(Krausse, 1999:67).

There is some evidence of a military presence (mercenary, auxiliary or legionary) at some major sanctuary sites in France. At La Chaussée-Tirancourt, a camp of Gaulish auxiliaries may be indicated by finds of Roman military material, pottery, amphorae and coins dated to between 40-25BC (Reddé, 1996). Similar material of the same date has been found at Ribemont-sur-Ancre, including nails from Roman legionary boots, and at Etrun. At Ribemont, it is suggested that an auxiliary unit may have wanted to benefit from using a site already dedicated to the gods of war, and "en même temps, ériger un temple plutôt tourné vers Rome et ses divinités protectrices" (Brunaux, 2000). Further excavations at Ribemont have pointed towards a drastic clearance of the site around 30BC (when, in the opinion of the excavator, the ruins of the previous features would still have been visible) followed by the erection of a new temple and sanctuary, which was immediately highly decorated, an unusual feature for this period. The evidence suggests an army construction of this building, as all the material found is military in type, and closely dated by coins. Brunaux (1999b) notes the continuity of the martial aspect of the site. It is possible that by replacing one form of sanctuary with another, the Roman authorities were underlining both their physical and their psychological conquest of the region. It seems apparent that the riches of the sanctuary sites were regarded as inviolate by warring Gaulish tribes - the treasures remained intact until taken by Caesar (Haselgrove, 1984).

The final correlate to consider is that of subsequent territorial demarcation and development. In this context, the Roman administration system reorganised the tribal lands into civitas capitals and their *territoria*, with or without regard to earlier frontiers, according to the patterns of alliance developed during the war and to the geographical location of the local capital. A study of the boundaries

"implies that the location of the administrative centres, rather than the territories of the eponymous groups, was the main determinant in the formation of the civitates. What determined the civitas to which the population of a particular area belonged was the nearest administrative centre. This make sense. The city in the Roman world was the civitas, and its citizens dervied their identity from this rather than from the territory they inhabited"

(Haselgrove, 1990a:62).

Over time, the tribal affiliations of the populations would have become blurred by the effects of the *civitas* administration, and older loyalties and ties would have been lost.

Conclusion

This case study has demonstrated many of the problems involved in any attempt to identify the symptoms of war in all its stages. Both Hamilton (1996) and Haselgrove (1990a) have pointed out the lack of functional evidence for the conquest of northern Gaul; the causal and preparatory stages of the war were virtually non-existent due to both the rapidity of the campaign and the fact that it was originally engendered by political events in Italy, rather than by local conditions. This leaves just the resultative correlates, which are surrounded by the usual problems of a lack of close dating, scarcity of finds and sites which have been well excavated, and confusion resulting from other forms of contact and influence such as trade and, perhaps, diplomatic alliances and gifts. A number of current and planned excavation projects (for example, the Late Iron Age fortified site at Olloy/Viroin in Belgium) may help to improve understanding in the future.

The conclusion that must be reached is that while resultative correlates can be identified, it would be almost impossible to ascribe these to warfare without the existence of Roman historical confirmation.

"Most of the other alleged evidence of the Gallic war turns out, on closer inspection, to be a question of more or less hypothesis, and if pressed further very easily leads to circular arguments"

(Haselgrove, 2007:513).

Case Study III - The Salt River Area, Arizona c1250-1450AD

Description

The area of this third case study is situated in East Central Arizona (*Figure 21*), with its southwestern corner at Phoenix, extending some 80-km north towards Perry Mesa and the headwaters of the Agua Fria River, and some 140km eastwards, the north east corner of the area meeting the cliff face of the Mogollon Rim just north of the source of Canyon Creek, and the southeast corner reaching the high ground between the Salt and San Carlos Rivers. This area has received an appreciable amount of study due to the large number of surviving sites and the concentration of major university archaeological departments in the Southwest USA.

The region is relatively high (around 6,000m above sea level at the edge of the Mogollon Rim and rarely dropping below 2,000m in the canyon bottoms) and is comprised of a series of deeply cut canyons between rocky mesa tops. Most rivers and creeks have a highly seasonal flow and there are a number of washes, dry except for the sometimes violent floods of the rainy season. For the most part, the canyon terraces and floors are narrow, offering only limited potential for agriculture, although there are a few wider valley floors, particularly at the confluence of the Salt and Colorado rivers, and in the lower reaches of the Tonto Basin. In these areas, irrigation canals have been identified and occasionally surveyed and excavated, although in the area of Roosevelt Lake, an artificial reservoir in the lower Tonto Basin, much evidence of early agriculture is now flooded. The topography is corrugated, often steep, and inhospitable, and the geology is based on upper volcanic strata, overlying older sedimentary rocks. Obsidian, used by prehistoric communities for toolmaking and the fashioning of mirrors, is found in a number of locations surrounding the study area.

The Hohokam core area of Central and Southern Arizona, bordering on the Mogollon and Ancestral Pueblo zones of Southern Utah, Southwestern Colorado, Western New Mexico, Southern Nevada and Northern Sonora and Chihuahua, enjoys its maximum period of precipitation in July/August (the moisture originating in the Gulf of Mexico) with a secondary maximum in December to March, originating in the Pacific. The modern annual average precipitation is about 20 cm. in the desert region and about 50 cm. in the uplands (Cordell, 2000).

The current land use is sparse, with very few settlements, only the urban sprawl of the city of Phoenix affecting any part of the study area. Much of the landscape is unused or only minimally exploited for agriculture or mineral extraction; other areas exist under federally determined legislation, such as the region designated as the Tonto National Monument. Vegetation consists of desert scrub, which nevertheless contains many species traditionally exploited by Native American peoples, including scrub oak (*Quercus terbinella),* Mexican cliff tree (*Purshia stansburiana*), Green Mormon tree (*Ephedra viridis*), various species of *Yucca*, *Acacia* and Mesquite (*Prosopis glandulosa*), other fibre plants such as Sotol (*Dasylirion wheeleri*), and medicinal, herbal and/or fibre sources including Prickly Pear (*Figure 22),* Agave, Jojoba, Sagebrush, Tomatillo and Flat-top buckwheat *(Eriogonum fasciculatum*)(National Park Service, 2004).

Cultivated species in prehistoric and historical times include Maize (*Zea mays sp.),* squashes and gourds (Curcurbitae sp.) and beans (Cordell, 2000); available wild fauna for hunting include deer *(*Odocoileus sp.*)*, rabbits [jackrabbits *(*Lepus sp.*)* and cottontails *(*Sylvilagus sp.)], bighorn mountain sheep (*Ovis canadensis)* and antelope *(Antilocapra americana)* (Ezzo, 1993*)*. Other species perhaps available for exploitation for skins and other uses include wolves, mountain lions, porcupines and snakes, as well as a number of bird species. Traditionally, porcupine quills, skins and eagle and other feathers feature greatly in local and ritual costumes.

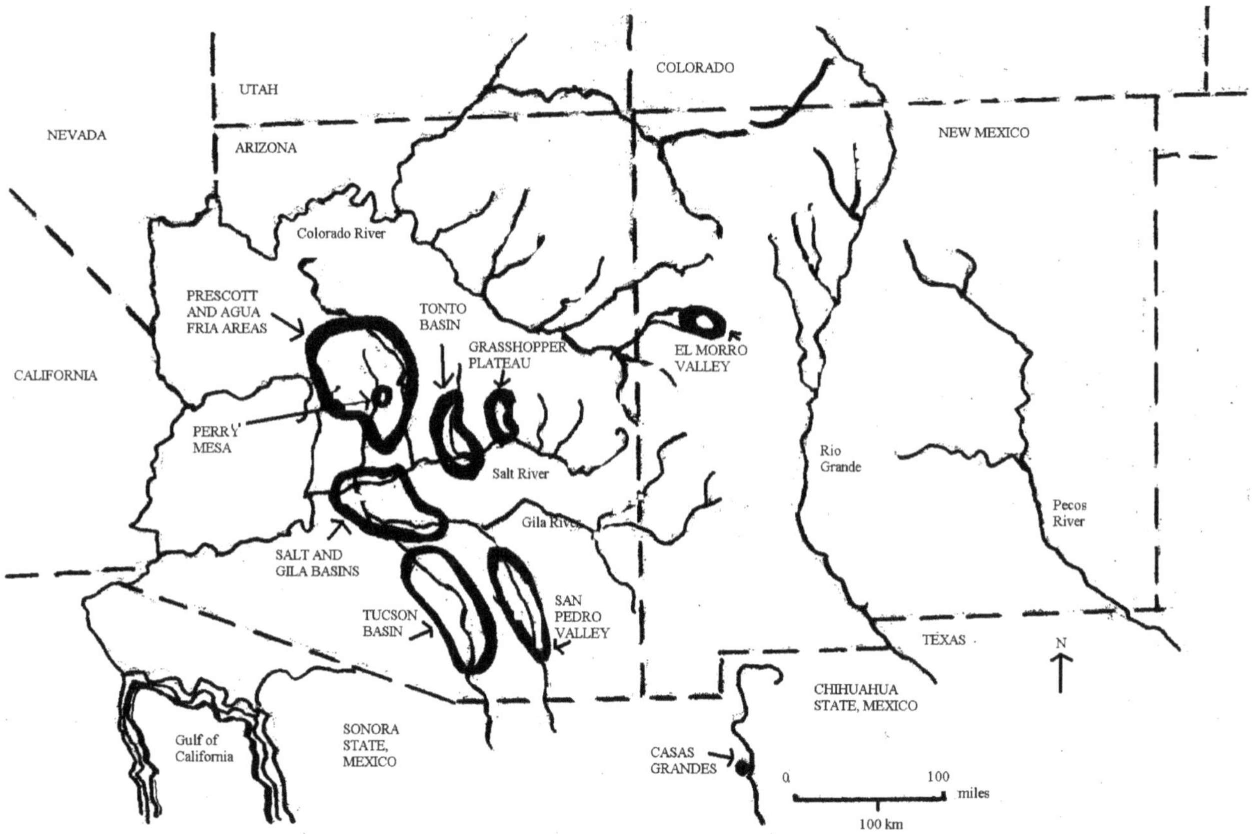

Figure 21: The Southwestern United States - areas of archaeological study (after Rice & LeBlanc 2001:3)

Figure 22: Typical vegetation in East-Central Arizona - Juniper and Prickly Pear (author's collection)

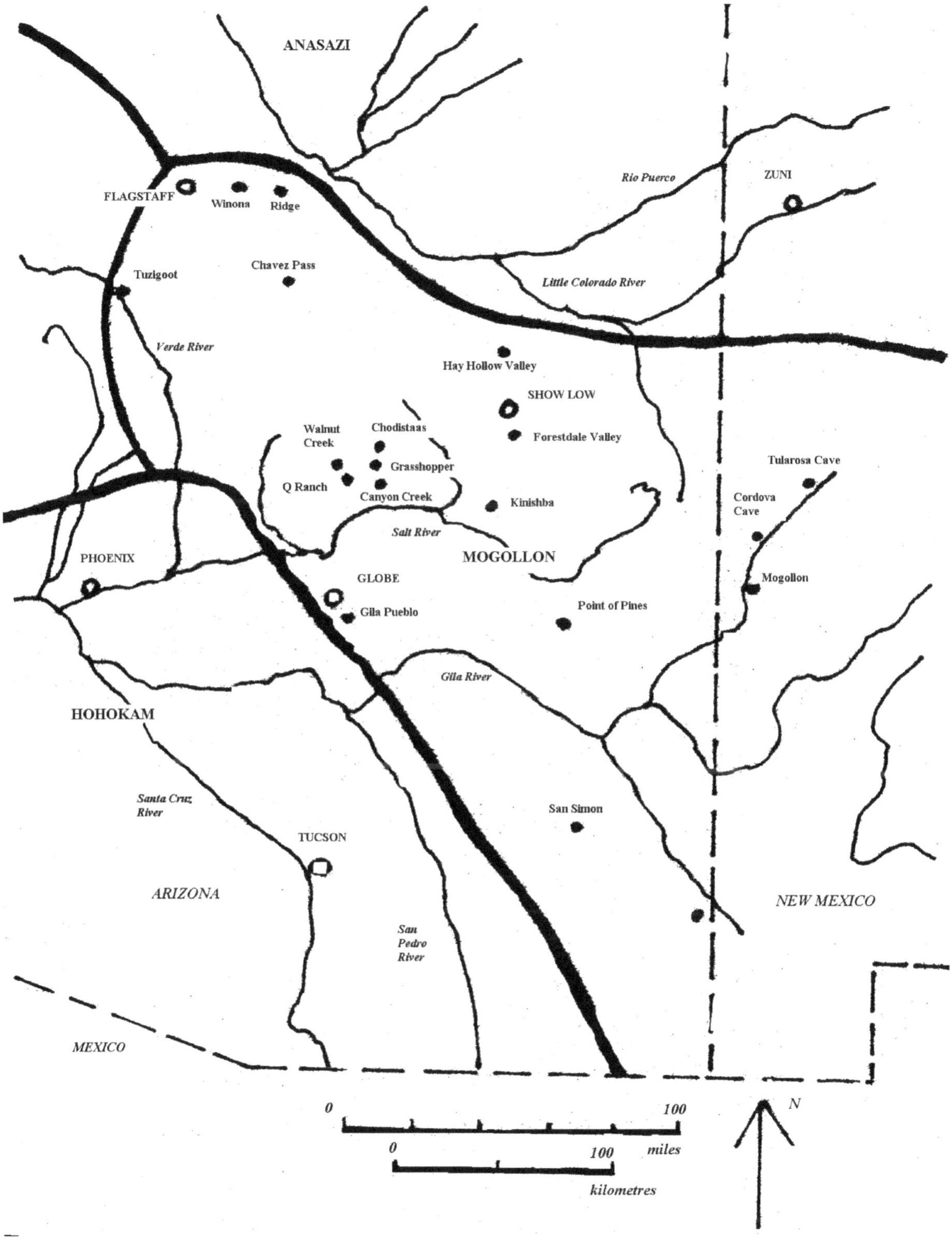

Figure 23: Cultural areas in prehistoric Arizona (after Reid, 1989:67)

Lithic tools attest to many centuries of nomadic hunting utilisation of the region, but more permanent settlement of the region began during the mid 6th century AD, with an ancestral culture known as 'Basketmaker'. By the 12th century AD, three major cultural groups (*Figure 23*) had developed in the American Southwest. In southwestern Colorado and northern New Mexico and Arizona, the Ancestral Pueblo (Anasazi) culture predominated, while in southern New Mexico and northern Chihuahua the Mogollon culture was ascendant. Central and southern Arizona and northern Sonora were within the Hohokam cultural zone (Hegmon, 2005). Differences between these three cultures are not highly visible except in pottery styles, some architectural detail and in some trade and exchange

PUEBLOAN	REGIONAL	LOCAL
Pueblo 1 c AD700-900	Early Classic	
Pueblo II c AD900-1140		Early Bonito Phase (AD900-1040)
	Middle Classic	Classic Bonito Phase (AD1040-1100)
		Late Bonito Phase (AD1100-1140)
Pueblo III c AD1140-1290	Middle-Late	Roosevelt Phase (AD1280-1320)
Pueblo IV c AD1290-1450	Classic	Gila Phase (AD1320-1450)

Table 14: Dating systems for Puebloan cultures

patterns. There is a general overlapping of subsistence strategy, major architectural styles and material culture, with the Hohokam of the south of the region being the longest lived.

Hohokam ancestry is claimed by modern Hopi peoples. The study area is at the north-eastern fringe of the Hohokam area, and is therefore to some degree affected by Mogollon neighbours to the east and Ancestral Pueblo trends to the north.

By about 700AD, communities in the region had adopted pottery, and were engaged in settled agriculture in several valley and canyon bottoms. In the Phoenix area of the Salt River Basin, extensive irrigation canal systems were established, with a range of small settlements within and flanking the system. These communities traded between themselves and more widely, and were part of a much wider Meso-American zone of trade, architectural ideas and, possibly, religious belief. Features such as ballcourts, developed in pre-Mayan cultures in Mexico and Guatemala, appear in Hohokam villages. Long distance trade items found at these sites include turquoise from the Sierra Madre Occidentale, obsidian from sites in Arizona, Utah and New Mexico, and marine shells from the coasts of the Gulf of Mexico and the Pacific. The period from about 1100AD to 1300AD is generally accepted to be that of greatest population and sophistication in the cultures of the American southwest, and the area along the Salt River was, by 1300, the core of Hohokam activity (LeBlanc, 1999).

However, by the time of first Spanish contact, nearly every major settlement, and most small ones, in the region was completely deserted. The explanation for this rests on two sets of data, one climatic and the other archaeological. The archaeological interpretations are divided into those which regard the climatic changes to be the most influential feature in the desertions, and those which consider that warfare was as important or more important a reason.

The region offers unusual archaeological opportunities. The low density of post-contact and modern settlement and utilisation, and the remoteness and difficulty of access have resulted in remarkable survival of sites and evidence; the presence of several very good university departments in the region has provided a source of manpower, finance and interest for the archaeological investigation of a number of sites. However, the very difficult terrain must mean that there is still very much to discover; winter erosion and flash floods have destroyed some sites, or concealed them, and some modern projects have been detrimental to evidence survival (such as the building of Roosevelt Lake mentioned earlier).

The study will concentrate on four major settlement groups: Perry Mesa, the Salt River Basin, the Tonto Basin and Grasshopper Mesa, for which good archaeological data are available.

There are a number of regional phasing systems, which apply to Hohokam, Mogollon and Ancestral Pueblo sites, and the use of phasing nomenclature is somewhat confused by reference to different systems in neighbouring regions. The following represents a basic comparison of these dating systems:

The period before 1140AD was generally one of growth and expansion, followed by intensification during the Hohokam Middle to Late Classic Period (1150-1450AD) in several distinct places. In Early Classic Hohokam villages, ballcourts were a reasonably common feature, but they disappear by the 12th century (Waters & Ravesloot, 2001). Development at the sites within the study area varies in scale and pace throughout the later period, and these also vary from developments occurring in Ancestral Pueblo and Mogollon areas. Therefore a brief resumé of the chronological history and a description of local conditions of each of the four concentrations of sites (Perry Mesa, Salt River Basin, Tonto Basin and Grasshopper Mesa) will be helpful.

Perry Mesa

Perry Mesa (*Figure 24)* is an area north of Phoenix along the Agua Fria River, and sites are stretched in a roughly NW-SE line along a 65 km stretch between the modern towns of Prescott and Carefree. Many sites occupy defensive hilltop positions and have been dated originally between 1100 and 1250AD, at a time when irrigation systems to the south were being abandoned.

The Perry Mesa sites are separated from the Hohokam core area by a 'buffer zone' 25 miles wide (LeBlanc, 1999:257). Spoerl (1979, 1984) divided these sites into four categories - forts, retreats, hilltop habitations and hilltop centres. This classification has been modified by, among others, Austin

Figure 24: Perry Mesa

(1977, 2000), who identified the forts as look-out points - small sites on hilltops or buttes, defensible and/or provided with defensive walls. Retreats are regarded as sites with temporary functions, again sited on hills and often with massive walls, but relatively small occupation areas.

These two types of site are remote from arable land. Hilltop habitations also have defensive walls but have evidence for more permanent occupation, while hilltop centres are much larger sites. The site AZ T:4:8 has been identified as a hilltop centre, and comprises a separate walled 'citadel' with habitation rooms within a larger walled enclosure with evidence of further habitation areas and cultivation terraces (Wilcox et al, 2001). Sites tend to form into clusters. Evidence for violence has been identified in many of these sites - missing body parts, crushed skulls and other signs of trauma occur. Analysis of line-of-sight links and siting has led to hypotheses that sites were interconnected in groups, with forts/lookout points and refuges occurring on the periphery of hilltop habitation and centre site distributions (ibid). Hilltop centres and larger sites typically have more line-of-sight connections than other forms of settlement (ibid) and seem to relate to rough circles of connection of about 18km (11 miles). Wilcox et al (2001) make a rough estimate of 300-500 people per circle. One way to account for this population density is that the area received immigration in the earlier part of the period, but this suggestion remains unproven.

The Pueblo III pattern of defensive building may have been dictated by feuding or raiding between the hilltop centre populations and their allies or dependent settlements as well as against neighbouring groups. However, by 1250AD most sites had been abandoned. A gazetteer of Pueblo III sites around the Perry Mesa region is be found in Appendix B. By the later 13th century, the distribution pattern of settlement in the Perry Mesa area had changed (*figure 25*). New pueblos were constructed on the mesa top itself, each associated with several smaller sites, and each commanding the head of a canyon providing access to the plateau and the rivers and creeks below. The major groups of sites can be listed as shown in Table 19 below.

It is notable that many of the larger pueblos are sited in cul-de-sacs, with very limited access. There are a number of smaller groups on Black Mesa, sited between Larry and Squaw Creek, at Larry Canyon, at the junction of the Agua Fria, and Lousy Canyon, between Perry Tank and Lousy Canyon, and at Horseshoe Basin. These sites circle the mesa top, covering every approach point from the river lands below, and have generally been dated to the period 1300 to 1450AD. The size of the larger pueblo in each region must also be regarded in the light of the total number of rooms

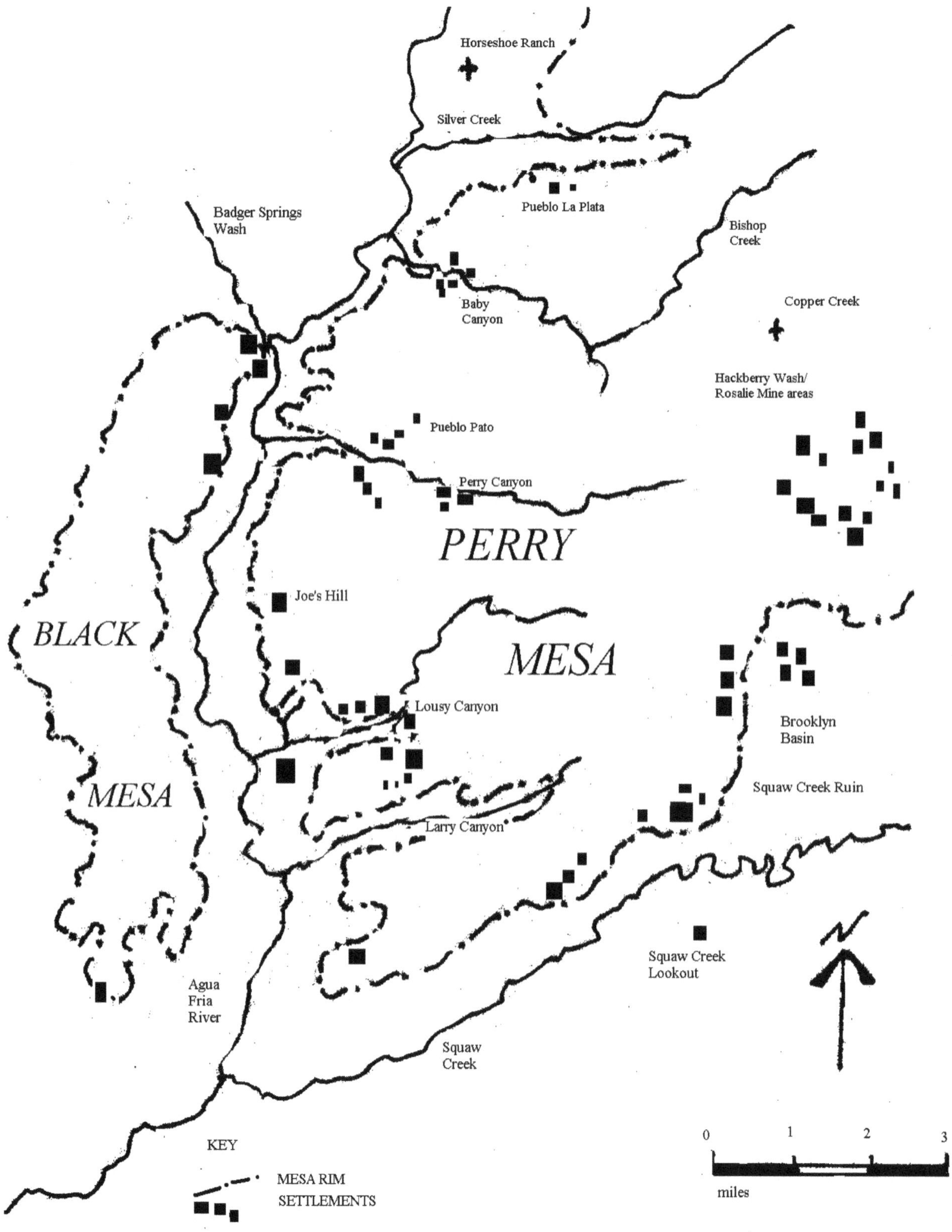

Figure 25: The Perry Mesa area (after: Wilcox et al. 2001:148)

in each system, to appreciate the possible size of population represented by each centre. A list of the Pueblo IV sites is presented in Table 20.

Population estimates are hypothetical only, as not all rooms are residential, many being used for storage and other purposes. The figures given below *(Table 21)* assume that 33% of rooms are not primarily residential, and that each

Brooklyn Basin Group	centred around	Site NA No. 13,472	70-90 rooms
Rosalie Mine/Hackberry Group	centred around	Site NA No. 13,477a,b	80+ rooms
Silver Creek Group	centred around	Pueblo La Plata	80 rooms
Bishop/Baby Creek Group	centred around	Baby Canyon Site	72 rooms
Perry Tank Canyon Group	centred around	Pueblo Pato	4 sites closely related Totalling c 410 rooms
Lousy Canyon Group	centred around	N:16:80	40-50 rooms
Squaw Creek Canyon	centred around	Squaw Creek Ruin	150 rooms

Table 15: Major groups of site in the Perry Mesa Area

*Table 16: Perry Mesa PIV Sites**

Brooklyn Basin			
Site NA No. 9869	50-65 rooms	4,000 sq.ft.	
Site NA No. 9870	5	4,080	
Site NA No. 9871	10	4,000	
Site NA No. 9872	10	3,820	
Site NA No. 9874	44	4,000	
Site NA No. 9875	37	4,040	
Site NA No. 13,472	70-90	4,020	
Site NA No. 13,473	20+	4,040	
Site NA No. 13,474	25+	4,090	
Site NA No. 13,475	25+	4,090	2 room blocks
Site NA No. 13,476	14+	4,090	2 room blocks
Site NA No. 10, 070 A-C	60+	3,860	3 room blocks
Site NA No. 25,989		5,510	walled enclosure
Rosalie Mine/Huckberry Wash			
Site unnumbered	39		
Site NA No. 10,087	20+	4,060	
Site NA No. 10,019	28	4,100	
Site NA No. 10,018	17	4,100	enclosed plaza
Site NA No. 10,020	31+	4,120	
Site NA No. 10,022	15	4,120	
Site NA No. 10,023	8	4,040	
Site NA No. 10,024	14+	4,040	
Site NA No. 10,065	32+	4,100	
Site unnumbered	7	4,290	
Site unnumbered	10	4,060	
Site NA No. 13,477a,b	80+		2 pueblos, 2 plazas, cul-de-sac
Site NA No. 13,478a,b		4,040	
Rosalie Lookout (25,989)	0	5,510	walled enclosure
Silver Creek			
Pueblo La Plata (11,648)	80	3,720	
Fort Silver	1+	3,600	wall on mesa tip
Horseshoe Butte			
Site NA No.25,985 5		4,232	walled enclosure, lookout
Bishop/Baby Creek			
Site NA No. 11,830	17+	3,610-60	
Site NA No. 11,604		3,750	
Baby Canyon Site (12,556)	72	3,600	enclosed plaza
Site NA No. 12,295	18-25	3,600	
N:16:46	30	3,700	cul-de-sac
Site NA No. 11,786	6	3,640	
Perry Tank Canyon			
Site NA No. 11,682	10-15	3,600	

Pueblo Pato (11,434A)	12+	3,650	
" (11,434B)	42+	3,650	
" (11,434C)	125+	3,650	
" (11,434D)	28+	3,650	cul-de-sac
Site NA No. 11,419	4-5	3,650	
Site NA No. 11,417	10+	3,640	
Site NA No. 11,418	9-12	3,680	
Site NA No. 11,438	9-12	3,760	
Site NA No. 11,440	5-8	3,760	
Site NA No. 11,490	48+	3,750	four pueblos
Rattlesnake House II (11,490)	48+	3,700	
Rattlesnake House I (11,439)	50	3,750	cul-de-sac
Site NA No. 11,785	9-12	3,730	
Site NA No. 13,327	8	3,700	
Between Perry Tank and Lousy Canyon			
Site unnumbered			fort
N:16:94	25	3,660	
N:16:95	9-12	3,800	
Site NA No. 11,792	35-50	3,650	
Lousy Canyon			
Site NA No. 11,452 (Joe's Site)	30+	3,650	
Site NA No. 11,420	3-4	3,480	
N:16:97	3	3,680	
N:16:96	30+	3,680	
N:16:80	40-50	3,690	cul-de-sac
N:16:90	4	3,700	
Site NA No. 13,303	9-12	3,690	
Site NA No. 13,329	5-8	3,760	
Site NA No. 13,350	5-8	3,520	fort
Site NA No. 13,317	38+	3,700	
N:16:12	15	3,620	
Near Junction of Agua Fria and Lousy Canyon			
Site NA No. 13,467	20+	2,520	
Larry Canyon			
N:16:6	8	3,740	
N:16:17	14	3,740	
N:16:16	9-12	3,700	
N:16:20	10	3,800	
N:16:57	7+	3,800	
N:16:59	1-3	3,840	
N:16:111	5-8	3,980	
Between Larry and Squaw Creek			
Site NA No. 13,469	40	3,320	
Squaw Creek Canyon			
Point Extreme Site (12,560A)	20	3,240	
" " (12,560B)	20	3,320	
N:16:54	10	3,300	
N:16:49	3-4	3,480	
N:16:50	3-4	3,480	
Site NA No. 11,436 & 13,466	20	3,840	
Squaw Creek Ruin (12,555 & 13,470)	150	4,000	large walled plaza; cul-de-sac
Site unnumbered	40		just north of Squaw Creek Ruin
Site unnumbered	c10	3,940	
Site NA No. 13,471	30	3,960	
Site unnumbered		3,640	fort; wall across mesa tip
Black Mesa			
Badger Springs Site (13,310)	40 (53)	3,500	plaza
Richinbar Site (B) (5423)	30-40	3,540	

* Variations in data reflect type and quality of recording; data recovered from mixed sources including aerial survey, ground survey, excavation and antiquarian records

Brooklyn Basin Group	370+ rooms	Population estimate	984
Rosalie Mine/Hackberry Wash Group	284++ rooms		688
Silver Creek Group	80 rooms		212
Horseshoe Butte	5 rooms	Temporary occupation/lookout	
Bishop/Baby Creek	137++ rooms		364
Perry Tank Canyon Group	417++ rooms		1,112
Between Perry Tank/Lousy Canyon	69+ rooms		184
Lousy Canyon Group	174++ rooms		464
Junction Agua Fria/Lousy Canyon	20+ rooms		50
Larry Canyon Group	54+ rooms		144
Between Larry and Squaw Creek	40 rooms		104
Squaw Creek Canyon Group	306+ rooms		816
Black Mesa Group	75+ rooms		200

(Data derived from Wilcox et al, 2001)

Table 17: Perry Mesa population estimates

residential room houses a family of an average of four individuals.

This would give a possible total population estimate of about 5,300 people living in an area about 80 square miles, although contemporaneity of room numbers and site use throughout the one and a half centuries of occupation cannot be established without uncertainty.

The siting of the settlements seems to be clearly based on defensibility. Vertical cliffs are supplemented by built walls, access routes zigzag and are delineated by blocking walls, natural crevices that might serve as access points have been blocked by placed boulders, sites are placed in excellent locations for lookout points with good line-of-sight relationships with residential locations, and approaches can be subjected to crossfire from paired outlying settlements. The mesa top itself has evidence of numerous small prehistoric fields and agriculturally related structures. This mesa is well placed to receivestorm systems coming north along the Agua Fria river, and is capped by a clay-rich series of volcanic and granitic soils in which basins of good soil and water can be built, collecting run off from the hilly topography. Rock-walled terraces, contour terraces, check dams, cleared fields and gardens have been recognised, although to date these have not been surveyed or mapped in any detail (Wilcox et al, 2001).

Abandonment of the Perry Mesa sites is thought to have been largely complete by 1450AD. The causes for that abandonment have been the subject of debate, but it seems clear in most accounts that warfare is recognised as a major factor, together with resource uncertainty; possible immigration of hostile groups may also be implicated.

Grasshopper Mesa

The Grasshopper region has been the subject of a field school research programme which ran from 1963 to 1992 (Tuggle & Reid, 2001). The area studied was 830 sq.km. between the Mogollon Rim and the Salt River, between Cibecue and Canyon Creeks *(figure 26)*. The mesa plateau rises from c500m high sheer cliffs from the valleys of the Salt and Canyon Creek. To the east, a series of terraces run down to the Cibecue. The north of the plateau is at about 1,900m asl, dropping to about 1,200m above the Salt River. The plateau is semi-arid but with a high water table and many springs. Game and wild flora abound, but agriculture is restricted to small suitable pockets of land with sufficient soil and water, and protection from frost. Crops were typically grown in small "garden" fields on the plateau and in limited terraced fields overlooking the Cibecue. The edges of the plateau are heavily dissected and rugged. Survey of 466 sq.km. of the plateau has resulted in the identification of 779 sites and 713 artefact scatters, while full-scale excavations have been carried out at four major sites: Grasshopper Pueblo, Chodistaas, Grasshopper Spring and AZ P:14:197, with a further site (Canyon Creek Pueblo cliff dwelling) having been investigated in the earlier 20th century.

The pattern of PII and early PIII settlement in the region is of small dispersed sites, with low population levels, and a subsistence economy in which hunting played a major role (Tuggle & Reid, 2001: Whittlesey et al, 1982). While communities round about began to grow in size and complexity, little change can be seen in the Grasshopper area until the late 1200s AD. Elsewhere, larger populations may have resulted from a greater proportion of agricultural reliance, and immigration following the Great Drought suffered throughout the Southwest United States in the middle 13th century AD.

At the end of the 13th century, small pueblos (or 'puebloid' settlements) of about 10-20 rooms began to be constructed close to water sources and arable land. These were usually in the form of room blocks *(figure 27)* and low-walled plazas. More isolated habitations and activity areas were centred on pueblos within a 1-2km diameter. Pottery studies suggest strong contacts with the peoples of the Colorado Plateau to the north, which was undergoing a

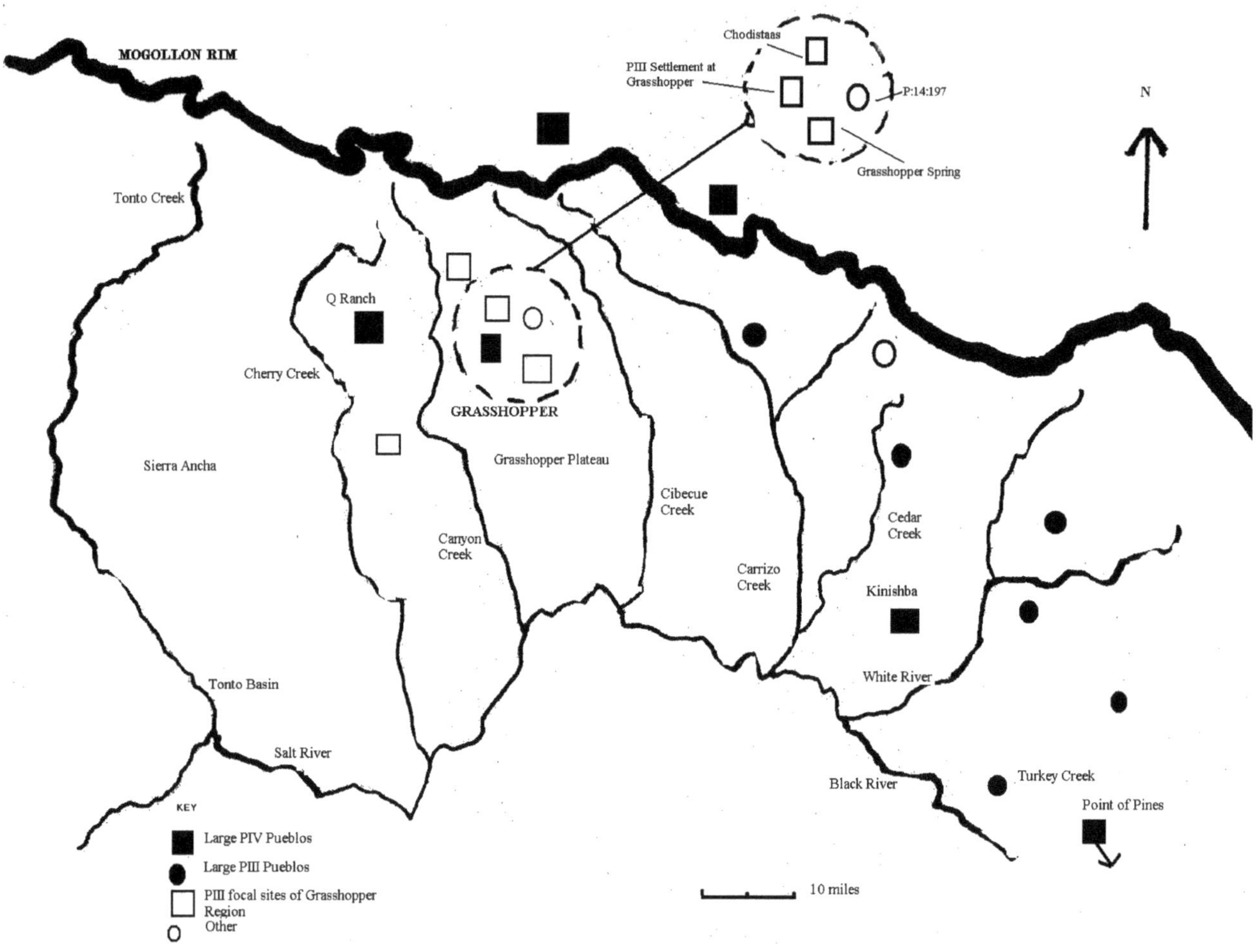

Figure 26: Surveyed PIII and PIV occupation sites in the Grasshopper Region (after: Tuggle & Reid, 2001:86)

period of extreme stress and population movement. Actual immigration is confirmed by isotope analysis:

> *"Strontium isotope analysis of the tooth enamel of 69 adults from Grasshopper Pueblo reveals aspects of the settlement of and immigration to the area during the late 13th and 14th centuries. Depending on the range of local strontium isotope compositions, non-local residents vary from one-third to more than half of the individuals analysed. Based on associated archaeological evidence, it is likely that the latter figure is more accurate. The majority of locals are associated with Room Block 2 and the Great Kiva. Individuals associated with Room Blocks 1 and 3 tend to be immigrant, and this is also the case of the two individuals analysed from Room Block 5, a room block which has been thought to represent an immigration tothe site. There are both locals and immigrants in the outliers, and the temporal data indicates that immigration continued throughout the occupation of the pueblo.*
>
> *Immigrants originate largely from two geologic areas: those underlain by Precambrian rocks located to the west and south of Grasshopper, or from areas underlain by Phanerozoic sedimentary rocks to the north and immediately east of Grasshopper. "*
>
> (Ezzo & Johnson 1997).

Conflict seems to have accompanied this migration in the late PIII. During this period, probably in the 1290s, Grasshopper Spring Pueblo, Chodistaas and AZ P:14:197 were all burned to the ground (Reid & Whittlesey, 1999:40). Also at Grasshopper Spring, a disarticulated body was found associated with a projectile point. Contemporaneously, use of the atlatl was completely abandoned in favour of bows and arrows. Immediately after these events, new room blocks were constructed at Grasshopper Pueblo, evidence suggesting that these housed people relocated from Chodistaas and Grasshopper Spring. Grasshopper Pueblo also has a more secure water supply than the earlier sites (LeBlanc, 1999). By the early 1300s, ten major aggregated pueblos of between 46-500 rooms had been erected on the Grasshopper Plateau, together with 58 small open sites and 24 cliff dwellings, some of which may have held permanent

	Phase:	No. rooms	
PIII Focal Sites			
Grasshopper Spring Pueblo	late PIII	15 rooms not defensively sited; burned	
Chodistaas	late PIII	18 rooms burned	
Glennbikii	late PIII	24 rooms	
Cibecue Snowflake Site	late PIII	24 rooms	
P:14:197	PIII	?	
Early Phase Grasshopper Pueblo	PIII	?	
PIV large pueblos			
Grasshopper Pueblo	PIV	500 rooms	enclosed plaza and great kiva
Brush Mountain Pueblo	PIV	150-200 rooms	
Blue House Pueblo	PIV	140 rooms	plaza; massive wall
*Canyon Creek**	mid PIV	58+ rooms}	1324-1346+
(cliff dwelling)		30+ rooms}	
Ruins Tank Pueblo	PIV	100 rooms	enclosed plaza; 2-3 stories
Spotted Mountain Pueblo	PIV	85 rooms	defensively sited
Black Mountain Pueblo	PIV	70 rooms	
Canyon Butte Pueblo	PIV	65 rooms	plaza; defensively sited
Hilltop Pueblo	PIV	65 rooms	plaza; defensively sited
Red Canyon Pueblo	PIV	45 rooms	plaza
Defensively located pueblos and sites - pueblos			
Oak Creek Pueblo	mid/late PIV	35 rooms	defensively sited guard s'ment
P:14:273	PIV	15 rooms	pueblo, cliff knoll guard s'ment
Shiprock	PIV	14 rooms	defensively sited guard s'ment
Pinnacle Pueblo	PIV	10 rooms	defensively sited guard s'ment
V:2:62	PIV	10 rooms	defensively sited guard s'ment
GFS 81-133	PIV	7 rooms	cliff/ridge end observ.post
GFS 80-56	PIV	6 rooms	cliff observation post
GFS 82-11	PIV	5 rooms	cliff/ridge end observ.post
Defensively located pueblos and sites - cobble habitations			
GFS 81-6	PIV	9 rooms	knoll observation post
V:2:73	PIV	7 rooms	knoll observation post
Masada II	PIV	7-15 rooms	cliff/knoll observation post
GFS 80-20	PIV	6 rooms	ridge end observation post
GFS 82-8	PIV	6 rooms	ridge end observation post
GFS 81-88	PIV	4 rooms	cliff/knoll observation post
GFS 82-12	PIV	3 rooms	cliff/knoll observation post
Selected cliff dwellings (*Canyon Creek included in large pueblos above)			
Red Rock House	PIV	30 rooms	cliff dwelling
Sanrace Cliff Dwelling	PIV	25 rooms	not defensively sited
Hole Canyon Cliff Dwelling	PIV	19-20 rooms	cliff dwelling
GFS 81-97	PIV	20 rooms	cliff dwelling
Double Springs Cliff Dwelling	PIV	13-15 rooms	cliff dwelling
V:2:78	PIV	15 rooms	cliff dwelling
GFS 80-52	PIV	10 rooms	cliff dwelling
GFS 80-60	PIV	10 rooms	cliff dwelling
V:2:63	PIV	9 rooms	cliff dwelling
V:2:75	PIV	7 rooms	cliff dwelling
GFS 80-64	PIV	5 rooms	cliff dwelling
GFS 80-61	PIV	3 rooms	cliff dwelling
GFS 80-63	PIV	2 rooms	cliff dwelling
Other settlements			
Cibecue Creek Ruin	mid/late PIV	c100 rooms	
P:14:281	PIV	31 rooms	
Salt Draw Pueblo	PIV	29 rooms	
GFS 76-162	PIV	24 rooms	
Lost Pueblo	PIV	19 rooms	
V:2:83	PIV	18 rooms	
GFS 81-95	PIV	16 rooms	
GFS 81-131	PIV	16 rooms	
Masada II	PIV	15 rooms	defensively sited
GFS 82-38	PIV	14 rooms	
GFS 82-16	PIV	13 rooms	
V:2:12	PIV	12-20 rooms	
Masada I	PIV	11 rooms defensively sited	

(Data derived from Tuggle & Reid, 2001: 106-107 and Wilcox, Robertson & Wood, 2001:190-191)

Table 18: Grasshopper Mesa PIII and PIV occupation sites

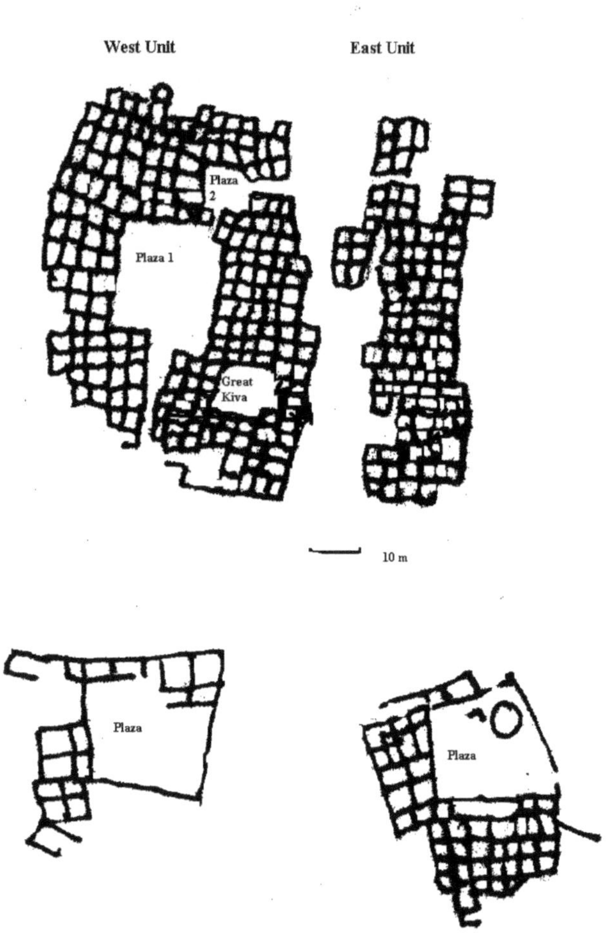

Figure 27: Room block arrangements at three sites in the Grasshopper region (after: Tuggle & Reid, 2001:92)

occupation. Population estimates for the Grasshopper region in the PIV era are:

1300-1330 est. number of rooms 1,276 population 2,041-2,552

1330-1375 est. number of rooms 1,729 population 2,766-3,458

(Wilcox et al, 2001).

> *"The cluster of the ten large PIV pueblos on the Grasshopper Plateau is not consistent with the patterning of agricultural potential in the larger area. The plateau is not a good agricultural locale and is certainly inferior to watered land available in the Cibecue Valley to the east and the Oak Creek/ Canyon Creek to the west, where, in contrast, there is only one small PIV site in each of these comparatively rich riverine locations"*

(Tuggle & Reid, 2001:94).

The large plateau-top pueblos are each situated in the best available agricultural locations, which are, however, limited. Along the rugged escarpment of the western plateau edge, there are many small isolated open pueblos and cliff dwellings, often placed on cliff edges with good viewsheds, and the limited amount of domestic detritus suggests temporary occupation. In some cases, access to these sites has been narrowed or blocked by boulders. It is considered likely that these were lookout posts.

> *"In addition to [these sites], there are two small pueblos on knolls at the edge of the plateau that were permanent villages but combined this function with defensive/observation positions. These are considered possible guard settlements"*

(Tuggle & Reid, 2001:96)

which had a role in warning the larger plateau pueblos of imminent danger. Some of the 25 western cliff dwellings may also have combined this role with storage and (semi) permanent occupation. Three small open pueblos on the plateau itself may also have acted as guard settlements, sited in isolated defensive positions. Pinnacle Point, on the eastern edge of the plateau, sits on a ridge which commands viewsheds of the Cibecue Valley terraces; to the west, Oak Creek Pueblo lies on a defensively sited rock outcrop above the valley but with relatively easy access to the plateau itself. The third site, Shiprock Pueblo, also sits on a rocky bluff, this time overlooking the Salt River. The combination of sitings results in a blocking of access routes right around the plateau (*figure 28*).

The larger sites are usually about 4 km apart, with the largest gap being 6 km; no study of intersite visibility has yet been undertaken. General abandonment, except at Grasshopper Pueblo itself, seems to have occurred in the 1350s.

The morphology of the larger pueblo sites themselves also shows a concern for defence. They are sited on high ground, on the edges of the plateau, or at the heads of canyons. Two storey room blocks present blank facades to the outside, and access to the residential complexes is via only a few narrow entrances. Only Grasshopper Pueblo is not defensively sited, but its size and architecture offer little opportunity for aggressive access. However, there is little evidence for deliberate burning at Grasshopper, and only three out of 674 individual burials show evidence of violence (scalping); there is an imbalance in the female to males population ratio of 3:2 (Reid & Whittlesey, 1999:206) which may relate to male deaths in distant localities during combat.

The main focus for defence seems to have been towards the south and west - the Salt River, possibly against the inhabitants of the Tonto Basin (who have also been suggested as aggressors for the Perry Mesa area). There is also the possibility of mutual aggression between the people of the Grasshopper Region and those of the Q Ranch system to the north west across Canyon Creek, where much burning has been identified. Q Ranch system guardposts and villages face their Grasshopper equivalents across the canyon.

Abandonment in the Grasshopper region appears to have been sporadic, over a period of between 30-50 years; the increased, aggregated populations were beyond the carrying capacity of the land, and there is evidence for dietary stress, a return to hunting and gathering, and a further period of drought after 1330AD. It has been noted that rooms abandoned late in the life of Grasshopper Pueblo usually contained much larger artefact assemblages than those abandoned earlier (Ciolek-Torello, 1985). Either the people were unwilling (or unfit) enough to contemplate transporting these goods to a new location, or a return was anticipated but never accomplished. It is thought possible that the population drifted southwards, but confirmation of this awaits analytical confirmation.

The Tonto Creek/Basin System

Tonto Creek runs towards the Salt River parallel and 50 km to the west of Canyon Creek. The creek forms a valley between 0.5 and 1.25 km wide, flanked by terraces which are wider on the east side of the valley than on the west (*figure 29*). The Lower Tonto area has been affected by the construction of a large reservoir, the Roosevelt Lake, which has obscured evidence of settlement over a stretch of land nearly 30km long and up to 3 km wide along the junction of Tonto Creek and the Salt River. In other respects, the geology, topography, flora and fauna are similar to those of the Grasshopper region. In the Tonto Basin area, some 143 sites have been surveyed and/or excavated. During the earlier part of the Pueblo IV period (Roosevelt Phase c1280-1320), settlement along the Tonto was dispersed; to the north, in the region of Cline Terrace, this consisted of small compounds, generally of no more than 10 rooms, and a smaller number of larger sites of up to 60 rooms. By the later part of the period (the Gila Phase c1320-1450) the 20 or so Roosevelt sites had shrunk to just five larger room blocks, with a few more isolated habitations and activity sites. In the south of the area, towards the confluence with the Salt River, a dispersed Roosevelt settlement pattern of 61 compounds and other sites is reduced to just one major site, Schoolhouse Point, with four other small isolated sites. Areas that had been settled for up to 400 years had been abandoned by the start of the Gila Phase. In total, for the region, 98 Roosevelt Phase sites, averaging less than 10 rooms each, developed into 13 Gila Phase sites, mostly with 26-60 rooms each; spacing between sites roughly triples in the Gila Phase to an average of 10 km between groups.

Gila Phase sites display features associated with defence and fortification; room blocks are built on hilltops, blocking walls and cliff dwellings appear, large compound walls are built (e.g. the 3m wide wall surrounding the Cline Terrace site) *(figure 30)* and defensive locations are selected. The morphology of sites alters in favour of compact block sites instead of open compounds, offering blank facades to the outside, and flat roofs allowing use as fighting platforms and as access to all parts of the settlement from within.

During the earlier PIV Roosevelt Phase, sites at Bass Point (overlooking the confluence of the Salt River and Tonto Creek) and Schoolhouse Point were relatively small, but both having evidence of ceremonial functions. At Schoolhouse Point, there was at least one large ceremonial room with a number of artefacts confirming this interpretation, including shell trumpets, stone batons, turquoise pendants and exotic plant materials, including tobacco (Simon & Gosser, 2001).

Most of the larger sites which lack specifically defensive siting are located in relatively high positions on the upper river terraces or on knolls (for example, Schoolhouse Rock, Indian Point and Rock Island).

> *"Less obvious forms of fortification may include the general shift to compact room blocks from compound-style architecture. Although these sites generally lack the large walled plaza spaces of compounds, they may have been easier to defend, as the exterior walls of the site are also the walls of rooms. Room walls may have been taller than plaza walls, were physically stronger...and provided a surface (the roof) where defenders could stand and strike down at attackers climbing the wall. Roofs on contiguous rooms offered a way for defenders to move rapidly from one area of the site to another as needed".*
>
> (Oliver, 2001:206-7).

		est. no. rooms	est. population
Upper Tonto Creek	1300-1330	340	544-680
	1330-1375	340	544-680
Lower Tonto Creek	1300-1330	697	1,115-1,394
	1330-1375	651	1,041-1,302
Tonto Basin	1300-1330	827	1,323-1,654
	1330-1375	475	760-950
		Sub-total for 1300-1330 = 2,982-3,728	
		Sub-total for 1330-1375 = 2,345-2,932	

(Wilcox et al, 2001)

Table 19: Population estimates for the Tonto Creek Area:

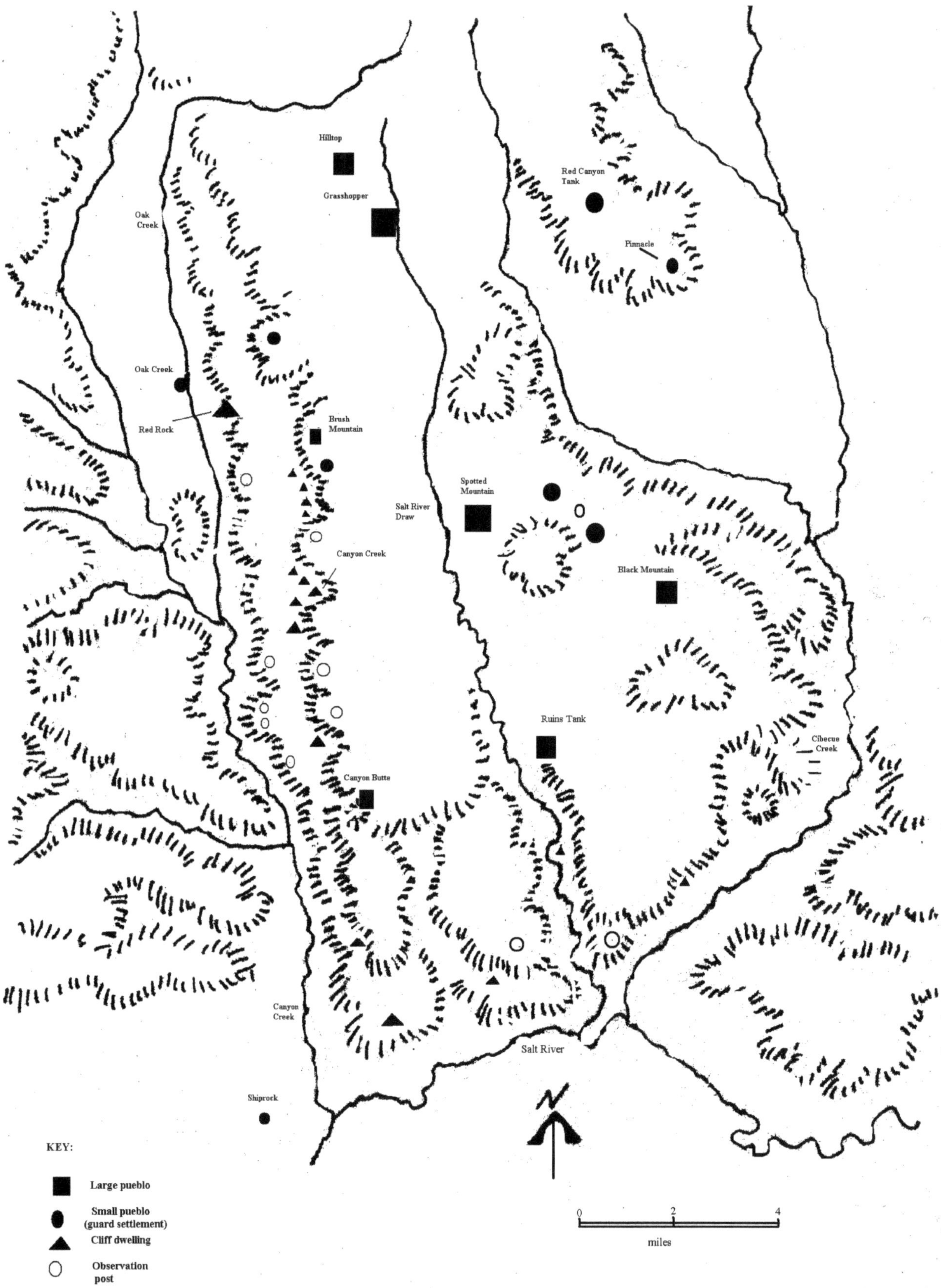

Figure 28: Grasshopper Mesa (after: Tuggle & Reid, 2001:88)

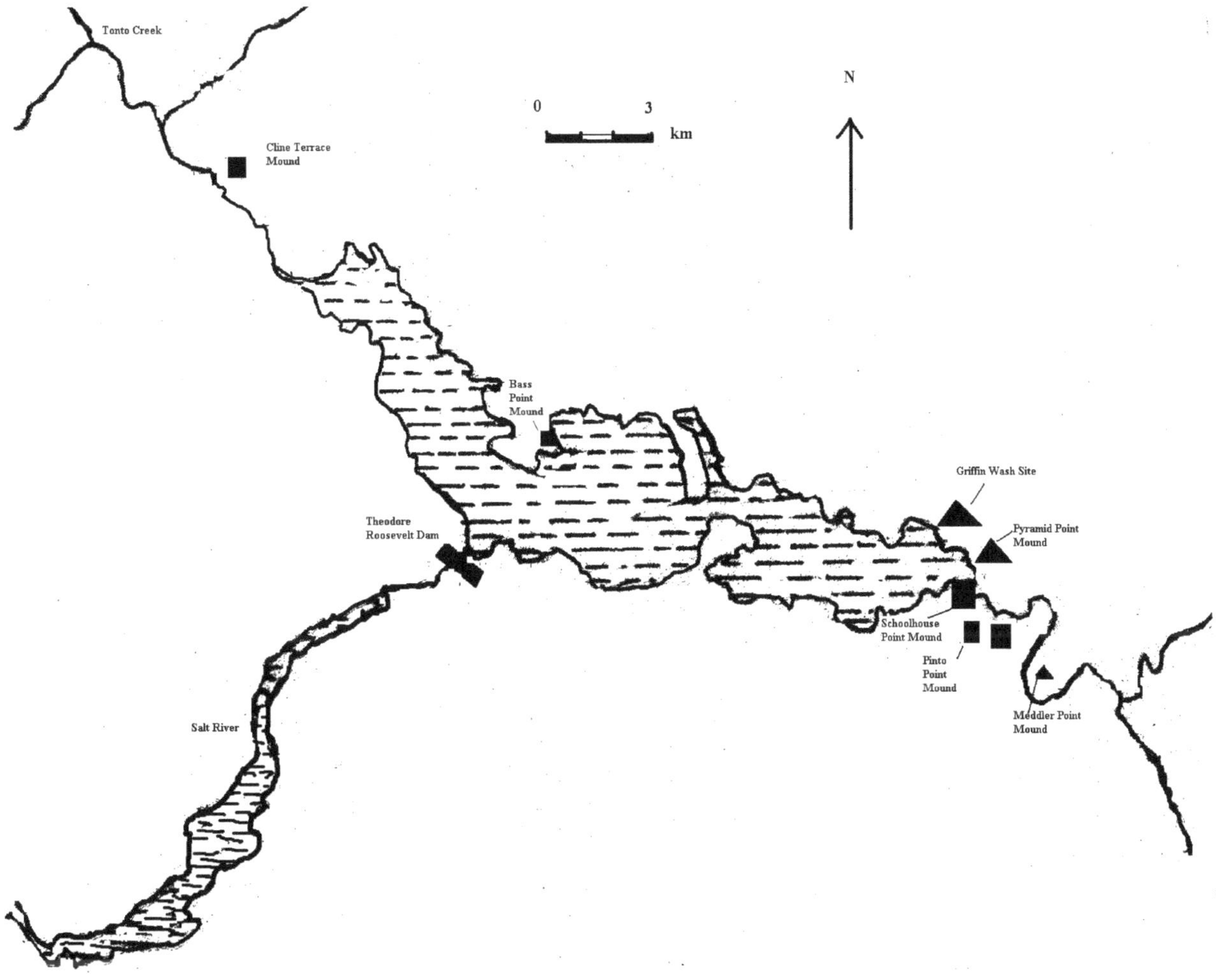

Figure 29: Tonto Creek and Basin area (after: Simon & Gosser, 2001:221)

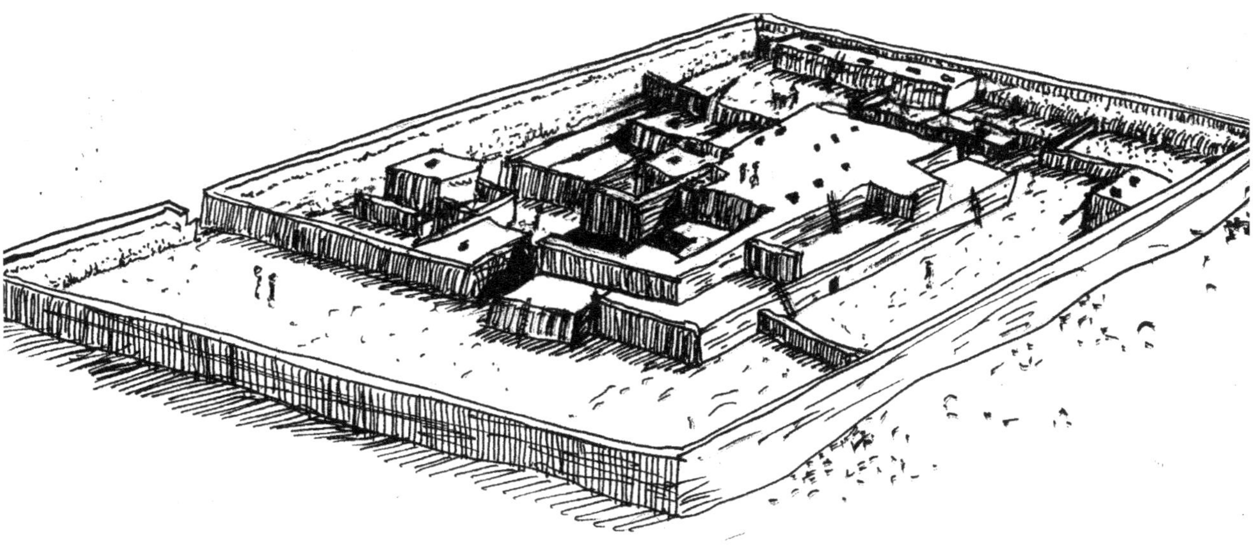

Figure 30: Cline Terrace Mound reconstruction (after: Simon & Gosser, 2001:225)

Upper Tonto Creek			
C.C. Griffin Ruin	undated	30-50 rooms	not defensively sited
The Box	undated	2 rooms	fort and compound
Black Mountain Ruin	undated	18 rooms	fort, defensively sited
AR-03-12-06-541	undated		lookout, defensively sited
AR-03-12-04-685	undated	none	lookout, defensively sited
AR-03-12-04-804	undated		fort, defensively sited
Rock and Rye Ruin	early PIII	c 40 rooms	
Gisela Schoolhouse	PIII	25+ rooms	platform mound, defensively sited
Lower Barnhardt Ruin	late PIII	50+ rooms	
Upper Barnhardt Ruin	late PIII	c20 rooms	
Rye Creek Ruin	late PIII/PIV	200 rooms	defensively sited, platform mound, tower
Houston Pocket Ruin	late PIII/PIV	35-40 rooms	defensively sited
Gisela Platform Mound	PIV	75+ rooms	not defensively sited
NA9747	PIV	30+ rooms	defensively sited
Lower Tonto Creek			
AR-03-12-06-581	undated	c 10 rooms	fort, defensively sited
Hamburg Ruin	undated	80+ rooms	not defensively sited
Trinity Ruin	undated	70-80 rooms	not defensively sited
Kayler Butte	PIII	48 rooms	fortified village, defensively sited
AR-03-12-06-809	PIII	80 rooms	
AR-03-12-06-810	PIII	14 rooms	
Haystack Butte Fort	late PIII	13+ rooms	
Park Creek	late PIII?	c 50 rooms	platform mound; not defensively sited
Horse Pasture	late PIII/early PIV?	c 30 rooms	platform mound
Chalk Springs Ruin	late PIII/early PIV	?	fort; defensively sited
Hackberry Basin Ruin	late PIII/early PIV	16+ rooms	defensively sited
Indian Point Ruin	late PIII/PIV	80 rooms	not defensively sited; burned
AR-03-12-06-295	late PIII/PIV	c 50 rooms	some burning
Cline Terrace	late PIII/PIV	36 rooms	platform mound; defensively sited; tower burned
Casa Bandolero	late PIII/PIV	50 rooms	
AR-03-12-06-769	late PIII/PIV	45 rooms	
Dresden Ruin	late PIII/PIV	70-80 rooms	not defensively sited
AR-03-12-06-2114	late PIII/PIV	20+ rooms	defensively sited; walled
VIV Platform Mound	PIII/PIV	c50 rooms	not defensively sited; burned
AR-03-12-06-576	PIV	20-40 rooms	
Eastern Tonto Basin			
AR-03-12-06-91	undated	3 rooms	circular compound
Las Sierras	late PIII	14 rooms	central plaza
Pyramid Point Ruin	late PIII	16 rooms	tower
Griffin Wash	late PIII	45+27 rooms	
AR-03-12-05-514	late PIII	13 rooms	defensively sited, walled
Tucker Box Ruin	late PIII	36 rooms	defensively sited
AR-03-12-06-2256	late PIII/early PIV?	25 rooms	not defensively sited
Conner Creek Ruin	late PIII/early PIV	62 rooms	defensively sited
Thompson Mesa Ruin	late PIII/early PIV	c 20 rooms	defensively sited
Pueblo Dinero Alta	late PIII/early PIV	50+ rooms	defensively sited
Meddler Platform Mound	late PIII/early PIV	15+ rooms	
Pinto Point Platform Mound	late PIII/early PIV		
Armer Ranch Platform Mnd	late PIII/early PIV	c 80 rooms	defensively sited; 2 plazas
Bass Point Platform Mound	late PIII/early PIV	15+ rooms	
Rock Island Platform Mnd	late PIII/early PIV	c 25 rooms	not defensively sited; walled
AR-03-12-06-2020	late PIII/early PIV	20+ rooms	fort; defensively sited
Bourke's Teocalli	late PIII/PIV	c 25 rooms	
Lower Ruin, Tonto Cliff	late PIII/PIV	66 rooms	defensively sited cliff dwelling
Upper Ruin, Tonto Cliff	late PIII/PIV	?	defensively sited cliff dwelling
Schoolhouse Point	late PIII/PIV	100+ rooms	
Tuzigoot on Salome	late PIII/PIV	100+ rooms	defensively sited
Armer Gulch Ranch	late PIII/PIV	34 rooms	burned
Porter Springs	PIII/PIV	50+ rooms	
Windy Hill Ruin	early PIV	40+ rooms	
Valentine Butte Ruin	PIV?	30-40 rooms	defensively sited
Hotel Ruin	PIV	50+ rooms	
Grapevine Ruin	PIV	20-30 rooms	

(Data derived from Wilcox et al, 2001)

Table 20: Selected Sites in the Tonto System

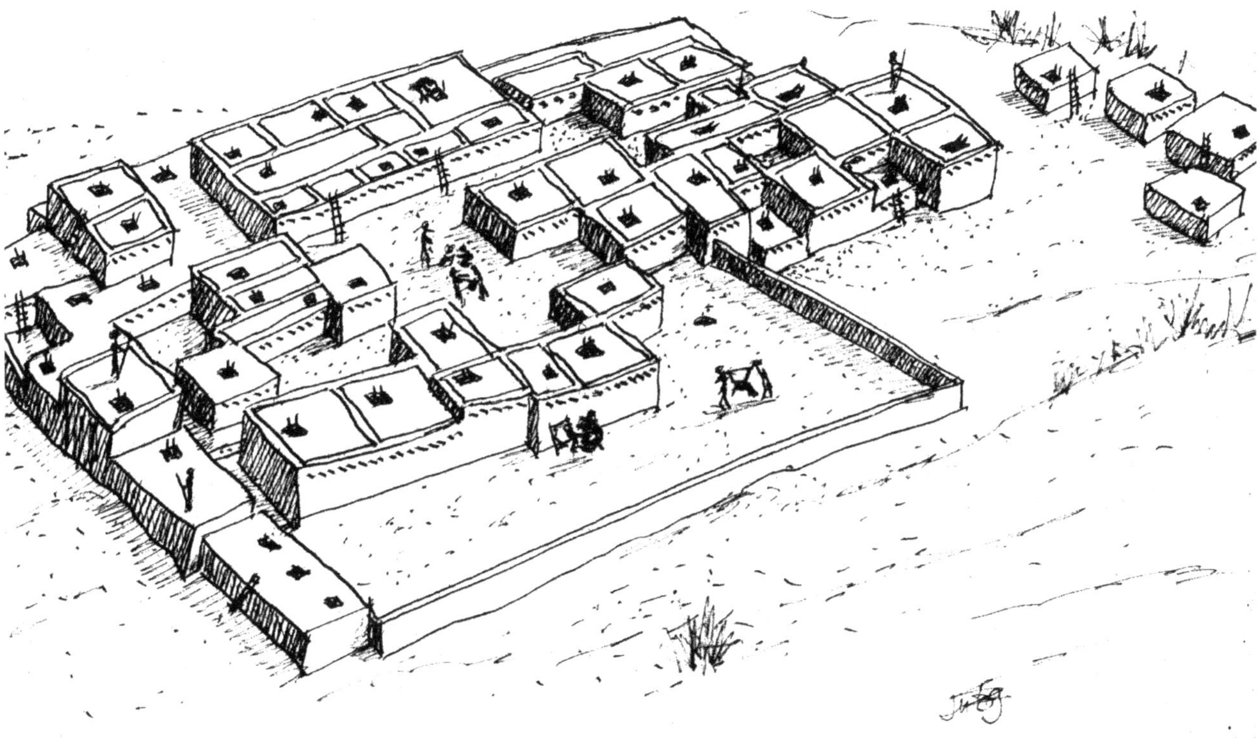

Figure 31: Schoolhouse Point Mound - reconstruction (after: Simon & Gosser, 2001:224)

By the Gila Phase, two large communities had developed, 25 km apart, in a landscape in which many smaller sites had become abandoned. Schoolhouse Point Mound *(figure 31)* was reconstructed employing elements of typical pueblo architecture; it seems to have been an aggregation of surrounding populations. The Cline Terrace Mound, however, "is primarily a structure of public architecture, with few residential rooms, surrounded by large public plazas" (Simon & Gosser, 2001:225). Unlike Schoolhouse Point, there were several smaller residential sites near Cline Terrace, which would have provided the necessary labour for construction.

There are a number of examples of skeletal trauma from the Tonto sites, as well as departures from normal ritual practice. Preservation of remains in the region is not particularly good, and has been affected by vandalism and grave-robbing, sometimes on a large scale. At a site called Grapevine Springs three male individuals were found lying beneath a collapsed roof, with evidence of burning.Similar remains associated with burned structures were identified at Schoolhouse Mesa, Schoolhouse Point and Ash Mesa. Burials not conforming to normal practice, and with evidence for peri-mortem dismemberment, have been found at Cline Mesa, Cline Terrace Mound and Schoolhouse Point; projectile points embedded in vertebrae have been noted from Schoolhouse Point and Cline Mesa.

Evidence for burning and abandonment has been studied in the Tonto Region. Cobble and adobe structures are not readily flammable in accidental circumstances, so evidence of destruction by fire is regarded as a good indicator of aggressive action. The presence of unrecovered artefact assemblages within rooms is also suggestive of destruction and abandonment during an aggressive period.

A study of 62 sites and 345 rooms returned results which show that 44% of rooms had been burned (151 incidences), and a large majority of rooms had unrecovered artefacts in situ (for example, at one site, of 22 rooms, 19 had been burned and 166 vessels were found in situ). The number of vessels left in situ in Gila Phase rooms is four times higher than that for Roosevelt Phase rooms. In the Roosevelt Phase, the total proportion of burned rooms was 36%, and for the Gila Phase, 82%. Attacks, therefore, seem to "have been dramatically more common during the later time period" (Oliver, 2001:214).

Evidence for burning as a single event, however, is hard to come by. At one example, a small Roosevelt habitation site known as AZ U:4:7, a large compound with a small number of rooms was accompanied by a large pit house in a central plaza, with a ramada-like structure against one wall. The ramada was filled with burned, and even vitrified, corn kernels, which had fused together, as were storage pits in the courtyard, and the remains of a shallow burial. The excavators consider that the burning of the ramada, the contents of the pits and the burial had all occurred during the same incident (Oliver, 2001:214). At a similar site (AZ:U:4:10), three conjoining sherds of a relatively rare effigy vessel were found under the burned remains of three rooms around a courtyard, again suggesting a single incident.

Artefacts found at the Tonto system sites also display changes during the earlier and later PIV period. Obsidian became more available in the Gila Phase, but was derived from more distant sources, most coming from Government Mountain, 185-200 km to the northwest. Turquoise and malachite are also more abundant in the later phase, oddly more common at sites furthest from sources. Marine shell artefacts declined in number throughout the PIV period, and are concentrated at those sites with ceremonial indicators. Exotic ceramics seem to indicate that during the Gila Phase, a major division in trade contacts developed at either end of the Tonto Basin, with Cline Terrace producing more evidence for contacts with the north west (Hopi yellow wares) and southeast (Hohokam brown wares) while at Schoolhouse Point, types from the Colorado Plateau in the northeast were more frequently represented (Cibola white ware and White Mountain red ware). Mogollon influences are more pronounced at Schoolhouse Point (Simon & Gosser, 2001).

It is possible that the two late concentrations at Schoolhouse Point and at Cline Terrace were in competition with each other, perhaps aggressively. It has been noted above that people from the Tonto System have been suggested as having also been in conflict with the communities of Perry Mesa and Grasshopper Mesa.

The Salt River Basin System

The final group of settlements to be considered in this study are located to the east of the city of Phoenix, grouped eastwards along the Salt River itself and down to the Gila River to the south. Much of the area has now been covered by urban sprawl, but has been the subject of a number of rescue excavations and surveys. The valley of the Salt River is much wider and flatter than the valleys of the creeks which form the river's tributaries, and it is surrounded, particularly to the south and west, by desert lands - the fringes of the Sonora Desert which stretches southwards down into Northern Mexico.

By 700-800AD the area had been colonised by communities which began to build irrigation canals, utilising headgate localities, of which there are about two dozen suitable sites along the Salt and Gila Rivers. Beyond these settlements, in the desert fringes, were a number of smaller communities who utilised springs, run-offs or shorter stretches of canal. The early pattern of settlement along the Salt River was dispersed, allowing the community to monitor the whole length of each canal. Larger settlements grew up on the ends of each canal group, possibly to defend access and protect the headgates. Evidence seems to suggest that conflict began as soon as a canal was long enough to carry two or more settlements, with control of water being a source of aggression from the start (Rice, 2001).

By the late twelfth century, climatic stress was affecting the peripheral desert communities, which began to relocate to defensive hilltop sites, or to fortify their settlements with walled compounds. By the early thirteenth century, these communities had been ousted from the regions surrounding the canal system settlements.

> *"The Hohokam irrigation system was by its very nature a catalyst for war. Hohokam populations living on large irrigation networks used war to manage the distribution of water through their canals, and at the same time the scale of their canal systems gave them considerable success in waging war against their less organised neighbours"*
>
> (Rice, 2001:305).

The canal system people grew much more numerous than their desert neighbours, and were more tightly bound into local alliance systems. Pottery exchange was stronger among communities along the same canal than with neighbouring canal groups, even when these were geographically closer Rice:2001).

During the PIII and PIV period, platform mound villages were constructed, the largest compounds being located at the ends of the canal systems: Casa Grande, Santan, Casa Blanca and Scottsdale.

At the beginning of the largest systems, Canal System Two, the settlement of La Ciudad was constructed along the edges of four parallel canals. It was a dispersed settlement, with groups of residences centred around cemeteries; each group consisted of between two and five free-standing wattle and daub structures built in shallow pits around a yard. These 'loci' are typically less than 500m apart, the intervening areas being utilised for horticulture. The loci associated with La Ciudad (*figure 33*) are, from north to south:

- *Los Solares, Belleview, Moreland, 21st Street and 22nd Street* along the northern canal;
- *Roosevelt* along the southern canal; and
- *Patrick and Fillmo*re between the Fillmore Canal and the Canal Ciudad.

The southern loci are later than the northern ones, which had already been abandoned by the time platform mound building had begun. Thirty-one per cent of the northern area shows evidence of burning. Each canal was about 4-5 m. wide and at least 1.5m deep, with the occupation areas within the system higher than the opposite banks

(Rice, 2001). Houses between the canals were protected by the canals themselves and by the proximity of other loci, but houses to the southwest, nearer to the external entrance to the system, were more vulnerable. Between 38 and 63% of the southwestern houses showed evidence of destruction by fire, compared to 13-25% of those in more protected

Figure 32: Surviving section of Hohokam irrigation canal, outside Phoenix (author's collection)

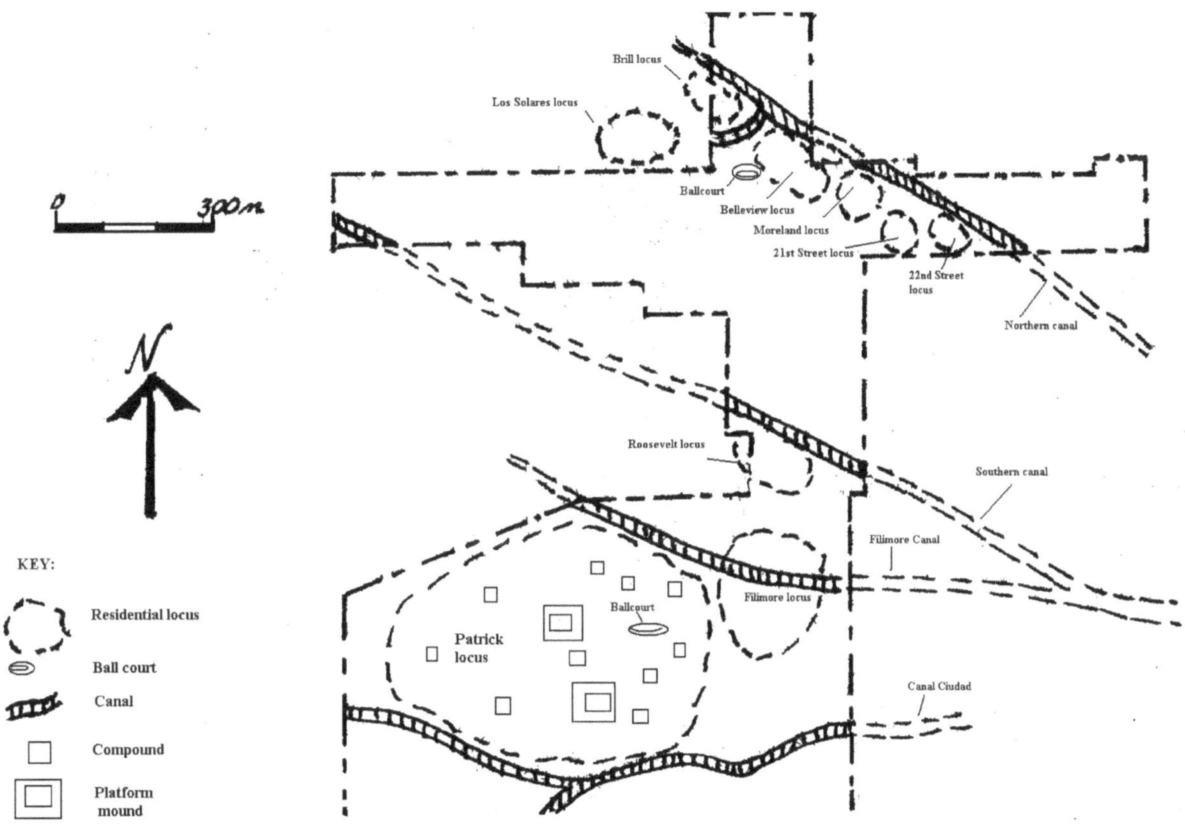

Figure *33: La Ciudad settlement loci (after: Rice, 2001:30)*

sites. A major change seems to have occurred around the end of the tenth century with the deliberate destruction of the ballcourt at the Belleview locus, followed by a period of about seventy years when levels of conflict seems to have declined. Towards the middle of the eleventh century, incidences of burning again rose, to about 50%, and by 1100 the northern loci had been abandoned (Rice, 2001).

Between 1200 and 1300, settlement remained dispersed and unfortified, but in the succeeding phase (1300-1380) walled compounds began to appear on the canal system, and there was depopulation of the surrounding desert areas. The site of Pueblo Grande, at the east end of the system, is representative of the new style of building. It consists of about 24 platform mounds within walled compounds (walls being about 2m high) over an area of about a kilometre. Each compound held between five and twenty rooms. It is suggested that the compounds could have protected stores and non-combatants during attacks, as well as providing a base for 'garrisons' of warriors (Rice, 2001). In some cases, outer walls are doubled, possibly supporting fighting platforms.

In the PIV period, a number of these compounds seemed to have supported elite residences; rooms built on top of the mounds display more elaborate construction methods and include chambers apparently designed for ceremonial purposes, with artefacts such as pigment bowls, shell tinklers, crystals and shrine stones associated with them (ibid).

The desert communities shifted to the sides and crests of high buttes surrounding the Salt River Basin, abandoning their valley floor settlements, a process that was complete by the 14th century. The new sites contained strong defensive features, such as corridor-entrances and loopholes commanding fields of fire in outer walls. At one such site, the Brazaletes Pueblo some 30km north of Canal System Two, excavation has uncovered a number of human remains displaying trauma.

Missing crania, hands and feet are recorded for nine individuals, crushed skulls for two, and in a further nine incidences, only partial body parts are represented. At least one interment is that of an elderly woman, and one is of a small child. Most of these burials were accompanied by grave goods - ceramics, bracelets, necklaces, shell tinklers, and projectile points. There is debate about whether these atypical interments represent the recovered bodies of warriors and other victims of war from the pueblo, or the remains of sacrificed captives (Wilcox et al, 2001). There is, as yet, no DNA or comparable data that can aid identification of the origins of these individuals, which in any case may not be sufficiently differentiated if warfare was relatively local, between neighbouring villages or settlement groups. The range of ages and the presence of both genders, together with the accompanying grave goods, might indicate that the former explanation is the more likely; sacrificial practices are only inferred for this region based on comparisons with Anasazi (Ancestral Pueblo) practices north of the Mogollon Rim. No hard evidence for such rituals among Hohokam and Mogollon communities within the area has been discovered to date.

Presence of causal correlates

Evidence for climatic change during the 13th and 14th centuries AD is strong, based on a number of sources including pollen analysis, faunal assemblages and dendrochronological data. The period 1276 to 1299 is often called 'The Great Drought'; precipitation levels fell drastically. From 1300 to 1330, the climate was unusually cool and wet, and then from 1330 to 1350 or so, a second, though less severe, drought occurred (Ezzo, 1993). Agriculture would have been badly affected by these events in this already marginal environment.

Number of years of:	1250-1274	1275-1299	1300-1324	1325-1349
mild drought	9	18	0	12
moderate drought	0	16	0	11
severe drought	0	4	0	3
stress for dry farming	12	23	3	24

(after Ezzo, 1993).

Table 21: Patterns of drought 1250-1350

Warming in the drought periods is also confirmed by the presence of *Peromyscus eremicus* (the cactus mouse) (Holbrook, 1982) and increased growth of desert scrub plants such as mesquite.

Demographic rise is also indicated in most regions of the study area. To some extent, this is due to an increased reliance on settled agriculture and an internal population expansion, but there is now evidence for migration into the region as well. In the Tonto Basin region, estimates for population in the period from 700AD to the mid/late 13th century suggest a steep rise in population between 800 - 900AD (growth rate estimated at 0.42%/yr), then more gentle rises of around 0.05% to 1275, followed by a fall in population to 1350 (Doelle, 2000). This trend is repeated over the Southwest generally, although decline in the southern part of the region is much more sudden and dramatic than in the north. The south shows higher population levels from 1200 to 1350 (peaking at about 100,000 around 1300-1325), and then a rapid fall to large scale abandonment and depopulation by 1450-1500. In the north, a maximum population level of just over 60,000 is reached around 1300, followed by a slower decline to a minimum of about 18,000 by 1700AD (Hill et al, 2004).

Immigration is suggested by a number of factors - ceramic change, architectural change, possible variation in burial ritual, and analysis of skeletal material. Strontium isotope analysis has identified 30 non-locals from 69 human samples from Grasshopper Pueblo (Price et al, 2002). The Pueblo has its origin in three separate groups of rooms, each possibly representing immigrants from the local region (from burned sites at Grasshopper Springs and at Chodistaas, possibly joining the local community), but these people would have shared similar strontium levels; the identified immigrants may have come from north or east of the Mogollon Rim, from the area dominated by the Ancestral Pueblo culture (Anasazi), which had been severely affected by the preceding decades of drought and by widespread conflict; Reid & Whittlesey further suggest that "an Anasazi enclave is represented....most of whom were women" (1997:158) and that this was an unopposed immigration as "Anasazi men joined the same ceremonial societies as Mogollon men, although none are present in the arrow society" (ibid). They suggest that the Anasazi women have married into the local population, but that certain ritual status, indicated by 'arrow society' membership, was reserved for the original male inhabitants. (Several such societies or 'sodalities' have been identified through analysis of grave goods; some were open only to males, others were represented in both male and female graves. The arrow society seems to have been male only, and associated with high status) (ibid:160).

The contention that different groups were present at the founding of Grasshopper Pueblo is further supported by "differences in architecture, in patterns of cranial deformation and burial practices, and in the distribution of ceramic types associated with the core units" (Graves et al, 1982). Architectural clues include the closely grouped room blocks, lack of mounds, compounds or plazas, and the insertion of the only Great Kiva late in the life of the pueblo. Adoption of cliff dwelling forms of settlement in the region is also a possible Ancestral Pueblo trait. Immigrants into the Tonto Basin in the Roosevelt phase may also have been from the Ancestral Pueblo area.

> *"There is a clear mixture of architectural styles and ceramics, suggesting that various groups occupied the basin. Some sites display a distinctly puebloan style of architecture that contrasts markedly with the typical compounds that were scattered across the landscape. Instead of growing larger by adding new, separate compounds, the sites marked by the pueblo type of construction were enlarged by adding rooms and compounds to existing features. The result is a larger number of contiguous rooms and less open space than in more typical compounds. A high percentage of corrugated pottery also was recovered from many of these sites, whereas the people living in compounds used more plain ware and red ware pottery"*
>
> (Reid & Whittlesey, 1997:253).

Climatic and conflict disruptions elsewhere in the region may have led to disruption in trading patterns and access to certain resources. At Chodistaas, occupied for 30-40 years in the late 13th century until it was totally burned and abandoned, use of high quality, high-fired Cibola Whitewares was very suddenly halted, in favour of more local and low-fired Roosevelt and White Mountain Redwares within a couple of years, around 1285, just before the destruction of the settlement in the 1290s. Use of the local redwares may reflect a desire for a less fuel costly technology generally, but may also represent a difficulty in accessing imported ceramic types.

There is no evidence for an increase in social complexity in most of the communities studied (e.g. Grasshopper Pueblo) (Upham & Plog, 1986). Except for sites in the Tonto Basin, habitations do not display any particular hierarchy or differential wealth or status. It has been suggested that political organization was based on an interaction between clan links and society membership, with power vested in particular families or clans who controlled or 'owned' specific important ceremonies. Most writers seem to agree on a largely corporate management pattern, with hierarchies developing only late in the period, and only in a few regions (Feinman et al, 2000). Elite platform mound residences are found in some of the Tonto Basin sites (as mentioned earlier), associated with ritual artefact assemblages, but otherwise there is little or no evidence for political change.

The landscape's morphology tends to create natural territories, divided by the rivers and deep canyons. There is no particular evidence for any large scale change in territorial organisation south of the Mogollon Rim during the period.

Presence of preparatory correlates

While a number of sites have been designated as 'forts', it does not seem that they operated in a way that could be designated as purely military. Actual fortifications are rare - the building of some linking walls is about the sum total of such structures. Nor is there a proliferation of weaponry, although this period sees the abandonment of the atlatl in favour of increased use of bows and arrows (Tuggle & Reid, 2001). Iconography related to warfare is represented by only one example, at Canyon Creek Ruin - a depiction of an individual with a shield. Such depictions are much more common in the Ancestral Pueblo region, but do not appear to have been repeated to the south even during periods of migration (Crotty, 2001).

Settlement abandonment, appearance of defensible settlement strategies, changes in settlement location and morphology, appearance of guard settlement and location of settlements related to line of sight links are almost universally part of the changes that occurred during the transition from the Pueblo III phase to the Pueblo IV phase and/or that from the Roosevelt to Gila phases. Perry Mesa has been described as a 'castle', with the settlements that ring the mesa top and control every access point taking

Figure 34: Cliff dwelling in the Tonto Creek area (author's collection)

the place of mural towers and barbicans (Wilcox et al, 2001:151). Similarly, aggregated settlement on Grasshopper Mesa rings the mesa top and controls access. In the Tonto Creek/Basin, new settlement aggregations are on defensible hilltop or terrace sites with good viewsheds. Settlement morphology changes in favour of blank exterior walls, rooftop fighting platforms and restricted access points. The communities of the Canal System Two placed large guard settlements at the ends of their canals and utilised the canals themselves as moated defences.

These changes are made even when the resulting locations are less convenient for access to farm lands or hunting areas - in the case of some of the cliff dwellings, the entrance to the sites was extremely difficult, and probably excluded access/egress to small children, the elderly or infirm unless assisted by more able-bodied members of the community *(figures 34 and 35)*. Care was taken to ensure that water and storage facilities were placed securely within the pueblos; for example at Canyon Creek Ruin.

Subsistence patterns also changed, although to what extent this was driven by climate rather than preparation for hostilities is unclear. Dry farming had faced major challenges during the Great Drought, but the move to mesa- and hill-top sites, and the abandonment of canal system farming in the Tonto Basin and valley terrace fields in the Cibecue Creek valley, seem to have occurred at the end of the drought, when the climate was becoming wetter. This certainly enabled mesa-top farming, at least in the short term, but it is hard to understand why the more established and regularly irrigated fields should also be abandoned. Certainly flash floods down some of the draws might have been a problem, but these are limited areas, and the floods would have irrigated lower ground in a usable manner. Mesa-top farming was limited to relatively small areas of suitable soil and water, and would, in the end, prove to be insufficient for the populations relying on it. Failure to utilise the former fields appears to be a maladaptive strategy, and must therefore be explained by reference to other factors, one of which may be warfare.

Changes in trading patterns have been alluded to in respect of pottery styles; much of the traded material, however, is likely to have been organic and little evidence survives. Unoccupied zones did develop, either between the large pueblo settlements or around the fringes of each group of

Figure 35: Narrow V-shaped entry into Tonto Creek cliff dwelling (author's collection)

sites, as occurred with the removal of the desert farming populations on the fringes of the Salt River canal systems.

Presence of functional correlates

Functional evidence for war is present in the form of burned sites and in examples of skeletal trauma. All three Pueblo III sites that have been excavated in the Grasshopper area were burned; burning occurred as a number of sites in the Tonto Basin and in some Salt River loci. Burned sites are also recorded at Lousy Canyon and Squaw Creek Ruin on Perry Mesa, and there is a record of a mass grave in this region (Wilcox et al, 2001). Many of the burned sites were not reoccupied, and there are particularly high levels of unrecovered artefact assemblages remaining in situ within burned structures - whole pots, metates (grinding stones), items of jewellery and tools. In a survey of 345 rooms in 62 Roosevelt and Gila phase sites in the Tonto Basin, 151 (44%) had been burned, and a total of 1,082 pottery vessels were found in situ - the greatest proportion of these coming from the burned rooms (Oliver, 2001:210-211).

Human remains excavated from Brazaletes Pueblo (a hilltop site occupied by people ousted from the desert fringes of the Salt River Canal System Two) display missing body parts, crushed skulls and casual burial within a room inside the settlement (Rice, 2001). At Grasshopper Pueblo there are few cases of trauma - just three incidences of scalping (Tuggle & Reid, 2001) - but Reid and Whittlesey (1997) have noted the 3:2 ratio of females to males and suggest that males may have been killed in warfare away from the pueblo. Data is unavailable to indicate how prevalent healed injuries may have been amongst the population; this information would be extremely important in any attempt to confirm that warfare was indeed implicated in these examples of trauma.

In the Tonto Basin a larger number of examples have been recorded *(Table 22)*.

It is noted by Oliver (2001:198) that bone survival in the Tonto area is generally poor, and that much material has been lost to vandalism. Many sites were excavated before modern procedures were adopted and evidence for skeletal injury, parry fractures etc., was not recorded. Oliver undertook a survey of the older records and identified some 129 individuals recovered from non-cemetery locations. These atypical depositions are overwhelmingly from the

Type of evidence	Site	Phase
Dismembered informal burial(s)	Cline Mesa	Roosevelt
Bodies in burned room	Grapevine Springs	-do-
Bodies in burned room	Schoolhouse Mesa	-do-
Bone tool in vertebra	Schoolhouse Point	-do-
Dismembered informal burial	Uplands Unit 3	-do-
Bodies in burned rooms	Ash Creek	Gila
Dismembered informal burial	Cline Mesa	-do-
Projectile point in vertebra	Cline Mesa	-do-
Dismembered informal	Cline Terrace	-do-
burial(s)	Mound	-do-
Dismembered informal	Schoolhouse	
burial(s)	Point	-do-
Bodies in burned room	Schoolhouse	-do-
Point		

after Oliver, 2001.

Table 22: Human remains data in the Tonto System

Gila phase, in a ratio of 5:1 to the Roosevelt phase. The likelihood of their being casualties of battle, however, is reduced by a number of factors. Nearly 50% were infants, and probably represent a tradition of interment of small children under house floors; 54 adult individuals were represented by only a few bones or teeth. It is possible that these remains were either disturbed from earlier burials during room construction activities, or were redeposited by scavenging animals. No clear indication of conflict trauma was recovered. It is also possible that dismemberment may have been part of a ritual or sacrificial activity practised in the region, whether with captives taken in war as the victims or with victims selected from amongst the local population.

Figures for normal cemetery populations are generally unavailable for most sites (due to opposition to excavation by local Native American groups, among other reasons). In other areas, such excavation has been possible. At Sand Canyon, Colorado, in an Ancestral Pueblo site, some 50% of the skeletal sample came from non-formal burial sites (Kuckelman et al, 2002). One context contained a mass burial of eight individuals, all displaying evidence of violent trauma which included cut marks to bone and crushing damage to crania. A hammer stone from the same site produced traces of human myoglobin, and similar traces were found on axes and in pots. These finds were associated with the 12th century drought period (ibid:508).

There is no record of weapons damage for any of the sites in the region - weapons were either of stone (projectile points) or wood (clubs), and therefore such evidence is not recoverable.

Resultative Correlates

The changes in settlement location certainly led to subsistence change, with a retreat from irrigation farming in most areas in favour of dry farming, which for the first part of the 14th century was possible in limited locations. There is also some indication that hunting activities were revived, with larger mammals becoming more common in faunal assemblages in the Pueblo IV period. This change, which is apparent at Grasshopper Pueblo and in sites north of the Mogollon Rim, has been explained in terms of the effects of aggregation and social change.

> *"...ungulate hunting may increase with aggregation, since one effect of aggregation can be to free hunters from the constraints of local agriculturalduties .... Pueblo IV peoples were willing to increase their long-distance hunting because of the importance of large mammals for community events...The importance of these events may have been the prestige and authority which individuals could gain...or the spread of rituals associated with the integration of aggregated communities from disparate backgrounds. The human population was shrinking at this time, not growing, so a population explosion did not prompt this increase in large mammal hunting"*

(Dean, R., 2001:283)

At Grasshopper Pueblo, the decades after 1300 were sufficiently benign in terms of climate to allow development of agriculture on the mesa-top; about 4 sq.km. of alluvial bottomland just south of the pueblo was cultivated, and the pueblo itself was sited at the north edge of this land at a point where water flow down the Salt Draw could be controlled (Graves et al, 1982). No effort was made to improve agricultural technology or to employ longer term water management strategies. However, by the mid 14th century (the period of population peak at approximately 600-1,000 individuals), pollen and faunal evidence points to a regeneration of woodland over abandoned fields in the vicinity of the pueblo, followed by a gradual abandonment (Holbrook, 1982). The years 1325-1350 would have been less successful, as a new period of drought conditions dominated, and although various alternative strategies may have been followed (including the splitting of the community to farm smaller sites at, for example, Canyon Creek Ruin and Red Rock House), the skeletal evidence suggests a general fall in dietary efficiency (Ezzo, 1993).

In the Tonto Creek/Basin area, a series of irrigation 'districts' have been identified with large areas of alluvial fan capable of supporting agriculture, and with varying areas suitable for canal irrigation. The region came under stress during the Great Drought, either because of the influx of immigrants from the north, or from the increased demands made on the irrigated land as other agriculture failed, or the result of both leading to increased erosion, lower flow levels and water tables and periodic flooding (Wood, 2000). Wood goes on to suggest that the building of platform mounds was a form of aggregation intended to manage parts of the irrigation system and to engage in competition (and perhaps conflict) with other similar communities in the area over water control. During the climatic amelioration of the early 14th century, and the move to higher, more defensive sites, the effort of maintaining the canal systems seems to have lost appeal, and parts of the irrigation systems were abandoned. Older systems of authority based on these also seem to have failed at this time. There is some evidence for dependence on a different suite of crops at higher altitudes (such as agave) until a return of drought conditions created a new crisis. By this time, the reorganization of labour and authority patterns was no longer able to adapt to climatic change, and the process of abandonment began, to be completed after flooding in the 1380s followed by another severe drought (ibid; also Waters & Ravesloot, 2001). Just to the north of the Tonto Creek sites, evidence from the Broken K and Carter Ranch settlements points towards a late period shift from the use of cultivars to an increase in exploitation of wild plant species and a reduction in hunting activities (fewer hunting tools but more grinding tools) as agricultural systems failed (Hill, 1972).

The puebloan aggregations maintained a level of long-distance trade, although not at great levels. At Grasshopper Pueblo, exotic trade items included macaws, copper bells, shell ornaments and ceramics (Reid, 1989), turquoise, obsidian, quartz, serpentine, gypsum and white-fronted parrots (Graves et al, 1982). Ceramics came from north of the Mogollon Rim, shell from the Pacific Coast, and the macaws and parrots from Chihuahua and Sonora in Mexico, possibly exchanged for agricultural surplus.

> *"Diverting some of the work force to manufacturing activities (even on a part-time basis) seems to have increased during the occupation span of Grasshopper, perhaps as a result of decreasing availability of new farmland coupled with the success of established trading enterprises. Other evidence suggests that the Grasshopper population increasingly engaged in trade as the local economy shifted away from agricultural self-sufficiency... Trade could have buffered the Grasshopper population from subsistence stress and at the same time sustained continued population growth.."*
>
> (Graves et al, 1982:200).

Changes in distribution patterns of ceramics have been discussed above with regard to the Tonto Basin sites.

No evidence for trophy collection has been found, unless we accept that the mutilated bodies discovered in some sites were prisoners of war, ritually sacrificed and buried in atypical manners within the settlements, and thus forming a type of trophy; nor is there any iconography other than the single example mentioned above. However, long bones possibly used for display have been identified at the contemporary high status Mogollon site of Paquimé in Chihuahua (Lekson, 2005).

Sex ratios at Grasshopper show an unusual preponderance of women in the cemetery populations [ratio of 3:2, or 82 male and 131 female individuals (Berry, 1985)] while elsewhere, ratios are more nearly balanced, but there is little hard data to work on in this respect - Tonto Basin remains are scarce and often in poor condition, and many of the Salt River sites have been destroyed by urban development before proper recording could take place. No data on Perry Mesa burials are available.

The gender imbalance at Grasshopper has been explained in terms of males dying while away at war (LeBlanc, 1999:235; Tuggle & Reid, 2001:102); Lowell (2007:104) suggests a different scenario:

> *"This imbalance was created not so much by Grasshopper men being killed away at war as by women and children [of whom there is an unbalanced proportion of between 65-68% each] migrating into the territory in larger numbers than men, because many of their men had been killed in conflicts elsewhere".*

Lowell further contends that this scenario explains Ezzo's (1993) observation of dietary difference - the immigrant women, lacking providing males, would have had less access to hunted meat and prime maize fields. Further, the fragmentation between room blocks at Grasshopper Pueblo suggests that Room Block 1, which contained more female and sub-adult burials, most evidence of food stress and less projectile points than the other domestic areas, points to a notionally segregated unit of these immigrants (Lowell, 2007).

The transition from Pueblo III to Pueblo IV resulted in the depopulation of large parts of the region, creating unpopulated areas between aggrandized pueblos, and around more stable communities such as those of Canal System Two on the Salt River. Fields and canal systems which had been used since about 700AD were abandoned across the region, although in many cases, this was probably due to climatic forces, with conflict being a possible contributory factor.

Almost total depopulation, abandonment and emigration from the region had occurred by the later 15th century, although the process was not uniform. Evidence from the Tonto Basin suggests that the first major phase of depopulation occurred shortly after 1300 (Hill et al, 2004:705) at a time when other sites were reaching a population maximum following agglomeration.

Certainly, by the Pueblo IV phase, a great deal of settlement relocation had taken place, and the morphology and siting of villages seems to have adopted a very defensive stance in many instances, even where this was maladaptive for subsistence. Aggregation of populations in the larger pueblos and compounds dictated a change to settlement morphology, with alterations in spacing (generally reducing free space from pre-aggregation levels), relationships of room blocks and the introduction of exotic forms of architecture and settlement planning.

The biodiversity of the region changed as a result of these movements, with the reduction of woodland in the new settlement areas in the building phases, development of new farmlands, and abandonment of older agricultural areas (Holbrook, 1982; Ezzo, 1993 etc.).

The maladaptive nature of the changes has been studied in the Grasshopper region (which offers the largest cemetery populations for analysis). The analysis of teeth from the population showed a high incidence of dental disease and tooth loss, as well as indication of childhood malnutrition and anaemia; there was also a very high infant mortality rate (56% of recovered bodies were aged under 12 years) (Reid & Whittlesey, 1999). Berry (1985:59) records a high percentage of osseous lesions and cranial hyperostosis, reflecting "rampant deficiency disease at the pueblo".

Cultural and ritual change is indicated in the later phases of some of the sites studied. At Grasshopper Pueblo, during a phase of expansion in the early decades of the 14th century, a Great Kiva (a ceremonial meeting room) was created by roofing in one of the plazas, that which was associated with the largest of the cultural groups represented at the pueblo, at a time which coincided with the renewal of drought conditions (Reid, 1989; Reid & Whittlesey, 1997). A habitation room was later converted into a kiva at Canyon Creek Ruin (Graves, 1983). In the Tonto Basin, platform mounds had been built in the Roosevelt phase, a type of settlement architecture which had not appeared in that area previously. However, there are no kivas or ballcourts in the Tonto area, suggesting that the inhabitants were either foreign to the general Hohokam and/or Mogollon cultures of the region generally, or there was a rejection of the cults represented by them (which are thought to have been largely concerned with the propitiation of rainfall and fertility) (Reid & Whittlesey, 1997). There is some evidence for a shift from cremation burials to inhumations following the drought, possibly due to a shortage of fuel (Waters & Ravesloot, 2001). It is in the later Pueblo IV period that non-standard burials appear in most sites - disarticulated bodies, bodies in shallow graves within living areas, and mass burials.

Territoriality certainly developed around the larger pueblos during the Pueblo IV period, with abandonment of smaller settlements and agricultural areas. For the Tonto Basin area at least, conflict between the larger pueblos in the system has been suggested (Reid & Whittlesey, 1997).

Conclusion

Most of the suggested correlates for war can be identified in East-Central Arizona between 1150 and 1450 AD, and many researchers have accepted the existence of warfare in the region during that period. Reference to more traditional indicators of war, as discussed in Chapter 1, are in fact relatively few - the defensive nature of building is inferred rather than formal in many instances; there are few instances of skeletal trauma, and those that have been found could be explained by ritual behaviour. Iconography to do with war is restricted to a single example, and there is no evidence related to weaponry that could not be interpreted as hunting equipment.

In fact, opposing theories have been advanced to explain the population aggregations of the later phases of the cultures of the region, and the subsequent abandonment of many settlements. These theories have been reviewed by Adler et al. (1996), and include (for aggregation) climate change, defence, agricultural and labour needs, alliance, status and exchange realignments, population density and the reduction of competition; for abandonment, causes include climate change, environmental degradation, disease, changes in land use and external relations, internal social breakdown, and warfare. Clearly, climatic factors were instrumental in driving necessary responses for survival; in this scenario, it might be supposed that co-operation would be more adaptive than conflict.The defensive siting of new settlements would seem to indicate that co-operative strategies were not being adopted. In the same conditions, reduction of competition would seem to be a positive step and may have been part of the motivation to combine villages into larger settlements and to create 'buffer zones' between them. The evidence of burned sites would seem to suggest that if this was the case, the strategy was not a complete success. In this marginal landscape, larger aggregations of population eventually proved to be beyond the carrying capacity of the environment; they were maladaptive and eventually failed. Aggregation as a means of dealing with labour requirements and the need to increase agricultural production cannot be supported by the evidence - the abandonment of some of the canal systems, where co-operative labour would have been essential, suggests that this was not a motive prevalent at the time. The siting of settlements in places where access to field systems was harder than before (such as cliff dwellings) also suggests that improvement in farming strategies was not a prime factor in decisions about site placement. The same arguments can be applied to the possibility of an increase in population density as a causal factor of aggregation; it would have been more useful to exploit a wider area to support the larger population than to abandon established fields. There is no evidence of effort in the Pueblo IV phase to improve farming efficiency or water management.

Where alliance, status and exchange relationships are a paramount concern, one might expect that some signs of elite control would come into being, particularly where people from different groups have combined together in an aggregated settlement. The evidence for elite groups is sparse, and where it does exist, it seems to refer to ritual and ceremonial behaviour rather than material wealth or status - room blocks have, in general, very similar assemblages of artefacts and forms of construction within each community.

Evidence for disease as a factor in abandonment is not, so far at least, recoverable in the region, and the pattern of burial does not suggest any type of devastating plague. In most instances, abandonment in the 15th century, although rapid in archaeological terms, probably took place over the span of several generations. Aggregation may have been driven, at least partially, by conflict; it does not seem that warfare played a major part in the final migration from the region.

Conclusions

This study has attempted to assess the usefulness of the production of a list of correlates as a tool for the recognition of the occurrence of warfare in prehistory, utilising three case studies to test the presence and validity of a proposed set of such correlates. The results from these case studies will be analysed in this section, and questions will be asked about the choice of correlates, the problems which have arisen from this choice, and whether or not applying correlates to the case studies has clarified or confused the contention that warfare may have occurred in some or all of these examples.

The case studies - results and discussions

The three case studies were chosen because each represents a different stage in the recognition and acceptance of the occurrence of war. In the case of the study of Gallia Belgica, there is historical confirmation for a period of intense hostility during the campaigns of Julius Caesar. The study of the Late Bronze Age in the Middle Thames area is of a period in which warfare might have occurred, but for which any confirmatory evidence is at best ambiguous, and at worst totally lacking. The case study of East Central Arizona represents a halfway point between these two - an area in which there is a general assumption in the literature that warfare took place, but for which archaeological confirmation may be questionable.

Table 27 illustrates the presence or absence of each correlate in each case study; these results are then discussed and assessed in more detail.

Causal Correlates

i) Climatic deterioration over the period

In the two prehistoric cases (LBA Middle Thames and East-Central Arizona) there is fairly good evidence for climatic change which may have precipitated change in society and settlement - drought in Arizona (Ezzo, 1993; Holbrook, 1982), cold and wet weather in southern England (Franzén & Larsson, 1998; Lamb et al, 1966; Parry, 1978 and others). In contrast, the period of the Julian campaigns in Gallia Belgica seems to have been one of fairly constant, even ameliorating, climatic conditions (Krause, 2003; Haselgrove, 1990b; Wightman, 1985). Given the probable stress created by changes where they occurred, particularly in the generally marginal agricultural conditions of Arizona, the case for considering climatic deterioration seems good. In both instances, subsistence change probably occurred, which in itself may have been a contributory factor to other social and political stresses. Climatic change may be relatively slow in human terms, occurring over several generations; the stresses on subsistence activities which are caused may therefore take time to become instrumental in creating the social and psychological conditions for aggression. The correlate is a strong one, but the timing of its effects may be variable and the degree to which aggression develops will probably depend on the scope and effectiveness of attempts at mitigation undertaken by the communities affected. The challenge, archaeologically, is in creating strong links between climate change and social changes taking place over possibly extended periods of time.

ii) Demographic rise

In all three cases, there is a general pattern of demographic rise, the result of centuries-long processes of agricultural intensification and specialisation. Figures in support of this rise are, however, ephemeral; it is in the frequency of settlement location and increased size of settlements that the evidence generally rests. The process of settlement proliferation and expansion in the region of Gallia Belgica has been noted and discussed by, among others, Krause, 2003, Bayard, 1996, Haselgrove 1996, 1990a, 1990b, and Wightman 1985. Work on Bronze Age settlement in the Thames Valley is less far advanced. However, in these two cases there is no indication of resource stress having begun to be part of the picture. It would appear that in the case of Arizona, the maximum carrying capacity of the land was being reached, exacerbating the stress created by climatic change (Hill et al, 2004; Doelle, 2000). Immigration from regions affected more drastically or, perhaps, slightly earlier, may have increased local problems (Reid & Whittlesey, 1997; Graves et al, 1982). Demographic rise has been postulated as a correlate for war by a number of theorists [including Haas (2001a), Ohlsson (1999), Ember & Ember (1997) and Vencl (1984)], but these case studies cannot confirm this correlate. However, where the correlation is visible, aggression may be more strongly indicated.

iv) Increase in social complexity

It can be suggested that the Late Bronze Age in Southern Britain was a period in which social complexity was increasing, with the appearance of elite centres (including Ram's Hill and Marshall's Hill) and large scale landscape reorganisation (demonstrated, among other features, by the 'linears' of Wessex) and the discrete groups of field systems (Yates, 1999). Late Iron Age Northern Gaul was also a far more developed social world than it had been - tribal groupings were, apparently, well recognised and confirmed by large oppida, coinage and alliances (Haselgrove 1996a; Roymans 1990; Wightman 1985). Conversely, there is little to suggest any rise of social complexity in 13th and 14th century Arizona. There are few elite centres (with the exception of Schoolhouse Point and Cline Terrace in the Tonto Basin) and no apparent differentiation in material culture or general architecture (Feinman et al, 2000). In general, the evidence is inconclusive to support this correlate in these cases, and in the case of the Late Bronze Age of Britain, it may be more useful to see warfare as a facilitator of increased complexity (as suggested by Webster, 1975), or social complexity arising from warfare (cf. Claessen, 2000) or to agree with Flannery (1972) that the evidence is too ambiguous to be of assistance. The case for this correlate is certainly not proven by the case studies.

v) Territoriality developing

Developing territoriality may be a resulting phenomenon of an increase in social complexity and/or of demographic rise. All three cases suggest some form of territoriality becoming more central to the organisation of communities. The 'linears' of Wessex have been mentioned; in the Thames Valley, the agglomeration of settlements and the apparent emptiness of the regions between these suggests a process that may involve an increase in group identity and, therefore, recognition of differences between each community. In Arizona, the Classic phase reorganisation of settlement patterns suggests the same processes at work. In Gallia Belgica, the tribal identity mentioned above seems to have been well developed (Roymans, 1990; Haselgrove, 1987). For Arizona, the separate mesa tops and canyons offer geographically well-marked frontiers; in the Thames Valley, boundaries are less visible, although the rivers and the apparent frontiers offered by the topographical changes from chalk uplands to riverine locations may be worth consideration (Field, 2001), and it is tempting to wonder whether the earthwork known as Grim's Ditch, which marks the frontier of upland/lowland pottery distributions, may be earlier than previously suspected. We can only presume, based on the records of Julius Caesar and other commentators, that the Gaulish tribes had a clear idea of their own boundaries, although these lack clear definition today. Certainly, incursions across boundaries can provide the instigating cause for aggression and socio-political strain between communities and some writers have suggested increasing territoriality as a causal factor for warfare [including Haas (2001a), and Jochim (1981)]. This correlate seems, therefore, to have some merit. The formalisation of territorial boundaries and increased effort in defence of these may be intrinsically connected to other forms of social and environmental stress, but whether identified as cause or effect will depend on better evidence and individual case histories.

vi) Changes to access for resources

Change to access for resources is also a strong candidate for consideration, although a complex one to consider. In the Late Bronze Age, the communities of the Thames Valley seem to have had a preferential access to high quality metal goods and goods imported from Continental Europe; the quantity of metal finds in the region is far in excess of any other part of the country. At the same time, artefact distribution relationships with Wessex and other parts of western Britain seem to have been minimal, with a clear pottery style-boundary along the line of the Berkshire Grim's Ditch (Barrett, 1980). It is possible that the emerging elite polities of the chalklands would have eyed this wealth with a certain amount of envy, particularly if they themselves were being more adversely affected by climate change than the riverine communities. The examples of Neolithic aggression in southwestern Britain are all associated with sites that have above average amounts of exotic artefacts (Carn Brea, Hembury etc.) (Mercer, 1981).

There is some evidence, particularly in ceramic distribution and different access to obsidian, that change in resources may have been a factor in Arizona, with the early collapse of the Ancestral Puebloan societies of the Colorado Plateau to the north (Graves et al, 1982), but the evidence is not particularly strong.

In Gaul, the effect of Roman trade imperialism has been cited as a major factor in the destabilisation of the native tribes and political systems, and Caesar used the closing of trade borders by the Belgic tribes as an excuse for invasion. The Remi tribe aided Caesar, providing him with a bridgehead from which to mount his assault against their neighbours, and thereby profiting from preferential trade deals with the Roman state (Haselgrove, 1990). It may be

	Case Study 1: The Middle Thames LBA (Warfare unproven)	Case Study 2: Gallia Belgica (Warfare attested historically)	Case Study 3: Eastern Arizona (Warfare suggested)
Causal Correlates			
i) Climatic deterioration over the period	Yes	No	Yes
ii) Demographic rise	Possible over previous centuries	No data	Yes
iii) Presence of other subsistence/social threats	No data	Yes (Roman expansion)	Possible - warfare to north of area
iv) Increase in social complexity	Probable	Probable	No evidence
v) Territoriality developing	Possible	Yes - tribal groupings and centres	Yes
vi) Changes to access for resources	Yes? Pottery/metalwork distributions	Yes - Roman effect on Gaulish trade	New sources for obsidian, pottery
Preparatory Correlates			
vii) Appearance of fortifications	Yes? Or status residences	No - already in place	Yes
viii) Proliferation of weaponry	Yes	No	No
ix) Iconography possibly related to warfare	No	No	No
x) Abandonment of settlements	Yes? Relocation to valley bottom?	No	Yes
xi) Appearance of defensible settlement strategies	Yes?	No	Yes
xii) Changes in settlement location	Yes - as above	No	Yes
xiii) Changes in settlement morphology	Yes?	No	Yes
xiv) Appearance of guard settlements	Possible	No	Yes
xvi) Change in subsistence patterns	Yes?	No	Yes
xvii) Change in trading patterns	Yes?	Yes	Yes
xviii) Unoccupied zones developing	Yes	No	Yes
Functional Correlates			
xix) Burned sites	No	No evidence	Yes
xx) Unburied bodies/ weapons trauma (human and animal)	No	No	Yes
xxi) Weapons damage	Yes	No	No
Resultative Correlates			
xxii) Change in subsistence patterns	Possible	Yes	Yes
xxiii)Change in trading patterns	Yes?	Yes	Yes?
xxiv) Trophy collection and memorials	No	No	No
xxv) Iconography possibly related to warfare	No	Yes (Roman)	No
xxvi) Deviation in sex ratio	No evidence	No evidence	Possible
xxvii) Depopulation	Yes at LBA/EIA transition	Yes - in places	Yes
xxviii) Unoccupied zones developing	Yes as above	Yes - temporarily	Yes
xxix) Population movement	Presumed	Yes - in places	Probable
xxx) Abandonment of developed arable	Yes	Yes	Yes
xxxii) Changes in settlement morphology	No data	Yes	Yes

	Case Study 1: The Middle Thames LBA (Warfare unproven)	Case Study 2: Gallia Belgica (Warfare attested historically)	Case Study 3: Eastern Arizona (Warfare suggested)
xxxiii) Changes in biodiversity due to human relocation	Yes - abandonment	Yes	Yes- abandonment
xxxiv) Changes in health profiles/standards of living	No data	Probable	Yes
xxxv) Artefactual and cultural change	Yes?	Yes	No
xxxvi) Changes in ritual and burial practices	No data	Yes	No data
xxxvii) Territoriality developing	No data	Yes (Roman administration)	No data

Table 23: Results of the case studies presence and absence of correlates

cynical to ask '*cui bono?*' when looking for the causes of aggression, but it can perhaps be argued that the motivation for economic gain is rarely that far removed from political decisions of this kind, particularly in cases where other stresses are being experienced. It is possible to speculate that other cases may provide stronger corroboration for this correlate.

Preparatory correlates

vii) Appearance of fortifications

The first correlate in this section relates to the appearance of fortifications. In retrospect, there is a problem with this - which comes first, the fortification or the war? Fortifications can be raised after the event (shutting the stable door after the horse has bolted). An example of this might be the extension of the medieval wall around Southampton only *after* the French had raided and burned the town in the fourteenth century. The archaeological problem is one of dating. Rarely are we in a position to be exact about the date at which a particular fortified site was erected, and warfare is often only clearly indicated if we can discern evidence of attacks specifically made upon the site. If the fortification has had the desired effect of deterring would-be attackers during a war, we are at a loss. e cannot be sure if the fortification was chicken or egg.

In the case studies, we have widely different scenarios. Fortifications do not appear in the Late Bronze Age Middle Thames with the exception of possible enclosures such as at the site at Marshall's Hill, Reading (Bradley, 1986) which is a presumed elite settlement, perhaps at Mayfield Farm, the palisade at Runnymede Bridge (Needham & Longley, 1986) and one or two other examples. In Gallia Belgica, the fortifications were pre-existing, relating to more extended periods of local conflict in the centuries before the Caesarian invasion, and there is no evidence for any attempt to add new sites at that period (nor indeed, was there probably much time in which to do so). It makes sense for a community under threat to look to re-use existing fortifications and there are many historical examples of this, such as re-occupation of Iron Age hillforts in the post-Roman period, which further complicates the dating problem. In Arizona, fortifications *do* appear - the building of walls, the stopping up of routes to mesa tops, and the building of look-out points that seem to have been only minimally occupied (in contrast to defensive residential sites) (e.g. Wilcox et al, 2001) but on the whole, these examples best fit the description "defensible settlement strategies" discussed below. The evidence in support of this correlate is extremely complex and fraught with difficulties of interpretation, so while it must be accepted that the presence of fortifications implies the *potential* for the occurrence of warfare at a particular place and time, it cannot confirm it.

viii) Proliferation of weaponry

The only case study to demonstrate a proliferation of weaponry is that of the Late Bronze Age in the Middle Thames, although it must be noted that most examples come from situations where a ritual (possibly funerary) context may be primary (York, 2002), in the form of deposition in east-flowing channels of the river. Deposition may also have occurred for purposes of status display and competition. In either case, weaponry may have had social value beyond its practical worth, and this in turn may suggest that warrior status may have held particular importance during the period. Little change is noted in the form or numbers of weapons observed in Arizona (Tuggle & Reid, 2001), and there is no evidence for any change in the provision or form of weaponry in Gallia Belgica at the time of the Julian conquest. Perhaps this should not surprise us too greatly. The dividing line between tools and weapons is hazy at best in prehistory, and where specific forms of weapon exist, it is likely that the eventual fate of these artefacts will not be archaeologically visible. Bronze weapons can be recycled, and the hoards of the Thames Valley (such as that from Petters Sports Field) contain many spears and sword parts (O'Connell, 1986). Iron weapons, if seized by an enemy, may be left to rust away, or taken away as trophies, or as part of a deliberate de-arming of a conquered population. Surviving pieces may be of particular heirloom value and so be finally deposited in other time periods and place, remote from their origin. Metal-poor communities would create most of their weapons from organic materials, and it is rare that these will survive into the archaeological record. Weaponry, therefore, must remain as a correlate of the *potential* for warfare but not as confirmation of its occurrence.

ix) Iconography possibly related to warfare

None of the case studies produced any iconography related to preparation for war. On reflection, this correlate is probably not useful and should be omitted from the list. However, iconography related to events during or after periods of aggression should still be considered and this topic will be discussed later.

x) Abandonment of settlements

xi) Appearance of defensible settlement strategies

xii) Changes in settlement location

xiii) Changes in settlement morphology

These four correlates must be considered together, as the evidence for each tends to overlap with the others, and to some extent, is inseparable.

Settlement abandonment as a defensive strategy is evident in the Arizona case study, where abandonment is also suggested as a result of climate change and, possibly, preliminary attack (for example at Grasshopper Springs and Chodistaas) (Tuggle & Reid, 2001). The canal-side settlements in the Tonto Basin relocated to mesa-edge or cliff sites, and while the climate may have been a factor, the choice of removal location can be seen to be maladaptive except in terms of defensive siting. The same maladaption has been noted for Grasshopper Pueblo (Tuggle & Reid, 2001). The location of the new settlements is clearly defensive, even to the extent of creating mobility and subsistence problems for the inhabitants. The siting of settlements around Perry Mesa has been likened to the building of a "castle" (Wilcox et al, 2001) with every access point restricted by walls or overlooked by buildings. In the Salt River canal system, the largest settlement agglomerations were located at the entrances to the system (Rice, 2001), providing protection for settlements further down the canals. The morphology of many of the Arizona settlements of the Pueblo IV phase does change, in some cases considerably. Defensive walls, large blank-faced room blocks and cliff dwellings all suggest a desire for increased security (Oliver, 2001; Rice, 2001).

The agglomeration of settlement into riverine settlements in the Later Bronze Age in the Thames Valley apparently involved the abandonment of settlements and ritual sites on the upper terraces of the valley at the very end of the Middle Bronze Age, as these have not produced any Late Bronze Age evidence. It is hard to explain these abandonments as being a result of any climate related incentive, as in a period of rising water levels and declining weather conditions, the better-drained terraces would appear to offer a more favourable location than the often flooded, always marshy bottom-lands. A defensive choice may therefore be a reasonable possibility as access to settlements was made difficult by marshy or flooded ground and narrow causeways between the inhabited gravel islands. A greater understanding of the conditions prevalent in the period would be necessary to confirm this interpretation, which is therefore only made tentatively. Flooding deposits and numerous drainage ditches and channels identified in those sites excavated do tend, however, to leave an impression of villages and farms that would have been difficult of access for much of the year, and with unusually damp and uncomfortable living conditions, an example of maladaptation if not defensively motivated. Agglomeration may also have been supportive of defensibility if it enabled warning systems (visual or audible) to be utilised for the protection of a large population, perhaps permitting a "massing strategy" for defence.

The morphology of the Thames Valley settlements does not appear to have undergone any major transformations from those of the Middle Bronze Age - they seem to be simply larger and more close together. However, it must be noted that none of the larger agglomerations has been excavated in its entirety.

No preparatory abandonment or other forms of settlement change are clearly visible in the case of Gallia Belgica, but instability may have led to a tendency for larger settlement groupings prior to the Gallic War (Bayard, 1996; Haselgrove, 1990b).

These correlates are considered to be useful and indicative, although complex and inter-related; the relocation of communities and the expenditure of effort in new building are not generally usual in agricultural communities unless there is a powerful incentive for change, and where such efforts may also be seen to have negative consequences for access to farming lands or for living conditions, a compelling need arises for a specific explanation for such changes. There is a difficulty in deciding to what extent such changes might be preparatory to or resultative from aggression - chicken or egg once again. The construction of sites in defensive locations or with defensive features may be indicative of a post-aggression reaction. There is not enough closely contextualised and phased evidence to be sure in the case studies.

xiv) Appearance of guard settlements

There appears to have been a deliberate siting of small settlements and lookout posts at a number of the Arizona sites, on mesa edges and at the points of access to the mesa tops (on Perry Mesa (Wilcox et al, 2001) and Grasshopper Mesa) and on prominent knolls overlooking wide swathes of country (particularly in the Tonto Basin (Oliver, 2001)).

The Thames Valley sites, being on low ground, lack the potential for easily identifiable guard settlements of a similar form. However, it is possible that some settlements, particularly those to the south and west of the main groups, may have fulfilled such a function. It is on the south and west that a few slightly more isolated, but enclosed, sites have been identified (such as Thorpe Lea), whereas on the north and east sides of the agglomerations, this type of settlement has not been identified. It must be stressed that the available evidence in support of this pattern is highly tenuous; however, if it should be confirmed, it offers an interesting possibility regarding the direction from which aggression was expected to come - namely the chalk uplands of the Wessex region. This correlate would seem, then, to be useful and worthy of further consideration.

In Gallia Belgica, the settlement patterns vary considerably across the region, but there is no evidence of guard settlements in the period before the Roman invasion.

xv) Settlement locations related to line-of-sight links

Detailed line-of-sight analysis has only been carried out for one area within the case studies - Perry Mesa (Wilcox et al, 2001). Here, strong correlation between settlement location and intervisibility is indicated. The definition of ancient line-of-sight links is fraught with difficulties - changes in land formation, use and flora make analysis difficult. In the Salt River area, the spread of modern Phoenix would make such studies almost impossible, and the Thames Valley has also seen a great deal of alteration due to building, quarrying and agricultural activities over the last 3,000 years.

In a relatively small area such as, for example, the Reading-Kennet group of Late Bronze Age settlements, there might have been an alternative strategy to visual signals. Here, the low elevation of settlements would have probably been a barrier to visual communications, but audio signals could, perhaps, have been possible, as the shape of the valley may have offered the acoustic properties necessary to enable horn calls or similar noises to carry for some distance. This form of warning system could have been particularly useful in view of the mistiness of these riverside locations at many times of the year. No scientific check has been carried out for this contention, which is based on personal observation at the location. Modern ambient noise in the vicinity of Heathrow and the M4 and M25 corridors would make such measurements extremely difficult. Much more work needs to be done on line-of-sight and other forms of communication analysis, but this remains a correlate with a possible usefulness.

Once again, no such settlement siting is evidence as a response to Roman aggression within Gallia Belgica.

xvi) Change in subsistence patterns

In the Late Bronze Age in England, there appears to have been a major change in subsistence in some areas, with the apparent slighting of arable fields and the construction of linear boundaries which have been associated with stock-raising (McOmish, Field & Brown, 2002). This phenomenon has variously been attributed to climate change, and socio-political change. For the people of the Thames Valley, there is certainly a shift away from arable agriculture - cereal remains are seldom found in excavations of Late Bronze Age sites, and the various ditches and banks visible alongside many of the settlements appear to be drove-ways. Traditionally, Thames-side pastures (up until the Second World War at least) have been prime lands for dairying and horticulture; it is possible to speculate that a similar subsistence pattern may have pertained in the Late Bronze Age. Certainly, the land in the immediate vicinity of many of the settlements is generally unsuitable for cereals, consisting substantially of heavy clays that are often waterlogged. For the same reason, cereal storage pits are not found in these areas.

Subsistence change in Arizona may also be indicated as communities moved up from the canal irrigated bottomlands in the Tonto Basin and the creeks surrounding Grasshopper Mesa. There may be some evidence, too, for changes in hunting patterns (Dean, 2001). In this case, the change may have been maladaptive, as the increased agglomerated population size may well have outstripped the carrying capacity of the land.

No major changes to subsistence patterns are visible in Gallia Belgica before the Roman invasion other than those associated with demographic rise and the general increase in settlement density.

Changes in subsistence patterns can indicate major shifts in social and political relations, given the inherent conservatism of rural communities, a conservatism related to the time and effort, as well as the long-term planning, involved in agricultural production. The *rate* of change may be crucial here; a rapid change suggests an anthropogenic cause, at least in part, while a change over several generations may have been more specifically climate or resource-depletion led, and far less obvious or traumatic for those involved. Thus this correlate would seem to depend on the ability to date, archaeologically, the process of change. In Arizona, the change seems to have occurred fairly rapidly - within 20-30 years. In Bronze Age Britain, the picture is less clear, as dating is insufficiently precise.

xvii) Change in trading patterns

The Thames Valley, as has been noted, profited during the Late Bronze Age from a great surge in trade in metal goods, more being found here than in any other part of the country. Many of these goods may have had either continental

European precedents or origins [such as brooches found at Runnymede Bridge, which have been identified by the site Finds Officer as Swiss (pers.comm.)]. There is also the development of an artefact boundary in the form of a fairly sharp division between Wessex pottery forms and those of the valley along the line of the Berkshire Grim's Ditch already mentioned (Bradley, 1980). The case of metal goods may be possibly interpreted as causal - the increasing metal wealth of the valley may have created the conditions for envy and/or aggression on the part of neighbouring communities. The pottery distribution may, however, be a symptom of a change in trading patterns away from older contacts, in favour of more local production, as such aggression began to be manifested.

In the case of Gallia Belgica, a change of trading patterns may also be discerned in the appearance of Roman goods, but in this case, historical evidence suggests that this was causal rather than preparatory. It was the rejection of Roman trade that, to some extent at least, apparently precipitated Caesar's attack on the region.

In Arizona, trade pattern change is visible at the time of the PIII/PIV transition - new (and more distant) sources of obsidian were exploited, more turquoise and malachite obtained, and ceramic studies indicate a schism in the Tonto Basin area, with different wares being adopted by different communities within the region (Simon & Gosser, 2001). These changes could be explained, possibly, if sites within or adjoining the region had become aggressive. This correlate, therefore, may also be usefully indicative, but it's use depends on relatively close dating and the identification of change agents.

xviii) Unoccupied zones developing

No unoccupied zones are visible in Gallia Belgica before the Julian period. In both of the prehistoric cases, apparently unoccupied zones are visible. The canyon floors and desert fringes of the Arizona settlement groups cease to be occupied (although not necessarily unexploited) in the late 12th century AD. In the Thames Valley, little settlement has been discovered in the areas between the major agglomerations (although again, this land may have continued in use) (Yates, 2007). In the relatively 'clean' archaeological environment of Arizona, this form of evidence seems to be irrefutable; in the palimpsest of the Thames Valley, there remains the problem of whether we have yet found all the sites of the period. However, in this heavily utilised part of the country, the pattern seems to be striking - the progress of development and, in particular, gravel extraction, in the region *should* have resulted in further sites coming to light if they existed. Where the agglomerations have been identified, there is no shortage of evidence, making the *lacunae* elsewhere more significant. Nevertheless, doubt must remain. Given sufficiently comprehensive study, this correlate could be useful.

Functional correlates

xix) Burned sites

Burned sites are not visible in the case of the Thames Valley sites. They are found in Arizona, and have been interpreted as indicators of warfare. All four areas studied in Arizona have produced evidence of burning episodes - all three excavated PIII sites on Grasshopper Mesa had been burned (Tuggle & Reid, 2001) and so were 44% of sites in the Tonto Basin (Oliver, 2001). Evidence from Gallia Belgica is almost entirely absent with the possible exception of Riemst-Lafelt (Bourgeois et al, 2003). The problems of using burned sites as a correlate include the identification of accidental or deliberate acts, with confirmatory evidence such as proportions of *in situ* artefacts or bodies, location of incidences of burning related to settlement morphology, and consideration of the way in which people deal with burned sites, all to be acquired. In the latter case, when structures are made of organic materials, there may be little evidence to be recovered - charred timber can be reused either as building material or as fuel, and sites can be scraped or washed clean for re-use. Burned daub or adobe readily breaks down in rain or frost unless buried, and fire-reddened earth may be interpreted as the result of industrial, ritual or feasting activities. Where a major settlement has experienced a significant amount of burning, has not been reoccupied, and conditions and materials are favourable for evidence survival, this correlate may be highly indicative; for many situations, however, it can be invisible.

xx) Unburied bodies/weapons trauma (human and animal)

Only in Arizona has this correlate been supported by any evidence, and even there, it is open to alternative interpretation as previously discussed. A number of factors combine to make this a difficult area to study (discussed in Chapter Two). It seems almost beyond belief that there is no evidence from Gallia Belgica, given the historical accounts of the Julian campaign, but clearly there must have been contemporary practices that have destroyed any evidence that existed. One may speculate that the Roman army utilised captives to collect and, probably, burn, victims of battles and reprisals in such a way that all trace has disappeared from the record. Burning and/or deposition of remains in fast-flowing rivers could explain the absence of trauma victims (and all ordinary deaths) in the Late Bronze Age of the Thames Valley. Unless the evidence is specific, therefore, [such as that found at Ribemont-sur-Ancre (Brunaux, 2000)] this correlate is often a poor indicator of warfare in ancient periods.

xxi) Weapons damage

The *use* of swords and spears in the Late Bronze Age Middle Thames Valley may be discernable (York, 2002; Needham, 1990) although the kind of damage found could result from practice fights rather than actual combat; the possibility of *preparedness* for war thus exists. Weaponry in Arizona in the 12th to 14th centuries may have been largely organic, or of stone which would not necessarily reveal identifiable use-damage. Once again, post-combat activities by the Roman army could explain the lack of weapons data from Gallia Belgica - collection and destruction of weapons would have been a sensible policy to prevent rebellion amongst the conquered tribes.

In many ways, therefore, this correlate, although historically popular in the debate, offers little in the way of confirmatory evidence for warfare unless, once again, the deposition is *specifically* contextualised (as perhaps at Hjortspring or Ribemont-sur-Ancre).

Resultative correlates

xxii) Change in subsistence practice

There is clear evidence of changes in subsistence practice following the Roman conquest of Gallia Belgica, in the form of new types of farm (many of which later developed into villa estates), new stock breeds (Kreuze, 1999; Lepetz, 1995), changes in vegetation profiles (including land clearance) (Kreuze, 1999; Defgnée & Munaut, 1995), farming practices (Lambot, 1996) and a redirection of production towards the needs of the occupying army on the Rhine. Change in farming practice is less visible in either of the prehistoric cases, if it is assumed that the changes noted above were already in place as preparatory phenomena.

Changes in the Late Bronze Age Thames region culminate in abandonments of settlements, but it is not possible, within the limits of existing data, to be certain in most cases that utilisation of the landscape for subsistence was also subject to either abandonment or change, with the possible exception of the field systems on the Heathrow terraces, which Yates (1999) considers do show evidence of abandonment. If, however, more land became uninhabited as a result of settlement agglomeration, the pattern of utilisation may also have altered, perhaps favouring stock-keeping as a preferred subsistence strategy. In Arizona, once people had moved to mesa tops and agglomerated settlements, there is evidence for change in farming practice including development of dry farming techniques (Graves et al, 1982) and the adoption of a new suite of crops (Wood, 2000). There is also some evidence for an increase in hunting (Dean, 2001). The eventual abandonment of both settlements and farming from the mid 14th century probably resulted from further periods of drought. The usefulness of this correlate once again depends on the dating of change and the specific context in which it takes place. Separating natural from anthropogenic or socio-political causes of abandonment is complex, and in many cases, evidence is insufficient to make a firm assignation of primary cause.

xxiii) Change in trading practice

The appearance of Romanised forms of pottery and other artefacts clearly indicates the presence of trade with the Mediterranean in the decades following the conquest of Gallia Belgica, although the persistence of some local forms reflects the probable poverty of the area and, perhaps, the strategy adopted by the Romans of allowing local leaders to continue to control their populations and some elements of production in the short term. A major change to trading patterns was created by the construction of the Roman road system in the region (Haselgrove, 1987); the arrival of Roman entrepreneurs is also a probability.

In Arizona, changes in pottery distribution suggest relatively sudden shifts in contact and manufacture (Montgomery & Reid, 1990) possibly as a result of fuel shortage brought on by the drought conditions, but also, conceivably, due to strained intra-regional relationships. Long distance trade continued, albeit at a low level, although it is suggested that at Grasshopper Pueblo, trade may have increased as a strategy to relieve the subsistence stress caused by agglomeration (Graves et al, 1982).

In the Thames Valley, there is a distinct localisation of pottery distribution developing from about 1000BC, with a clear break in distribution patterns more or less along the line of the Berkshire Grim's Ditch, marking the boundary between the Thames communities and those of the chalklands to the west; at the end of the Middle Bronze Age, Wessex types were widely found in the Thames Valley, but after this period they virtually disappear (Barrett, 1980). Thames forms are equally rare in Wessex sites of the later period.

Pottery is a relatively good indicator of trading contacts; less easy to recognise are patterns of change in the distribution of other commodities, but the studies also demonstrate changes in the acquisition of obsidian in Arizona (Simon & Gosser, 2001), and in the access to metal goods in the Thames Valley, which has much higher proportions of these artefacts than elsewhere in southern England in the Late Bronze Age. However, very little metalwork of any type from the Early Iron Age has been found in the study area, suggesting that with the adoption of the new material, the Thames valley trade links went into decline (Bradley, 1980). There are multiple possible reasons for these apparent changes, but their use as indicators of changed social and political climates must remain a possibility, and so this correlate may be useful.

xxiv) Trophy collection and memorials

xxv) Iconography possibly related to warfare

None of the case studies offer strong evidence for these two correlates with the exception of a few military tombstones in Gallia Belgica (Wightman, 1974). Such evidence is available for other places or periods, however - the site at Ribemont-sur-Ancre, the Hjortspring deposit, warlike iconography in the Ancestral Pueblo area north east of the Arizona study area at the same period, later Roman triumphal architecture and depictions in France and Germany are all relevant to the correlates, if not the case studies. However, depictions of military victories do not confirm military action - they may represent political aspirations or 'spin' as much as actual events. These correlates probably retain their usefulness in other contexts.

xxvi) Deviation in sex ratio

There are no data for sex ratios in the Late Bronze Age populations of the Thames Valley, nor for the post-Julian populations of Gallia Belgica. In the Arizona case, studies at Grasshopper Pueblo have shown an apparent preponderance of female inhabitants in the later periods of the site's occupation (Berry, 1985), and it has been suggested that this may represent either the loss of male inhabitants to conflict, or the arrival of refugees deprived of their menfolk, or the capture of women from enemy communities. Without more evidence, this correlate is weak; it is also subject to influences created by differential gender burial practices and standards of evidence recovery.

xxvii) Depopulation

xxviii) Unoccupied zones developing

xxix) Population movement

xxx) Abandonment of developed arable

These correlates may be regarded as interconnected and will be discussed together. Depopulation is evident in Arizona, where most sites have become abandoned by 1400AD (Rice & LeBlanc, 2001), and, apparently, in the Thames Valley, where survival of settlements into the Early Iron Age is apparently almost nil, although doubts about pottery dating may have skewed this conclusion (Yates, 1999). [Post-Deverel-Rimbury ware forms are currently subject to some uncertainty in dating in the Thames east of the Berkshire Grim's Ditch (J. Cotton, pers. comm.)]. In both these cases, there is an apparent increase of unoccupied areas during the period (with abandonment of fields associated with canal systems and river terraces in Arizona, and the agglomeration of settlements in the Thames Valley creating areas between settlement clusters which have produced remarkably little in evidence for use and occupation for the period). In both cases, there was population movement into larger settlements or groups of settlements. At the end of their respective periods, both the Arizona sites and the Thames Valley Late Bronze Age sites appear to have become abandoned, with populations having moved away almost entirely. In addition, there is possible evidence in Arizona, at Grasshopper Pueblo, of the presence of alien groups, Mogollon or Ancestral Puebloan, either as refugees or as captives (Lowell, 2007). There is little evidence to suggest where the people went at the end of the period, although it has been suggested that some, at least, moved further south to join other surviving Hohokam populations in the Tucson Basin (Adler et al, 1996).

The case of Gallia Belgica is more complex. If Caesar's account is to be believed, there *should* be evidence of depopulation; however, this is largely missing, with the possible exception of the area around Tongres and in the tribal territory of the Eburones, which do appear to have a period of abandonment before being resettled by peoples from beyond the Rhine, relocated as part of Roman policy. In these cases, a hiatus in occupation levels is apparent and subsequent re-occupation is accompanied by some changes in material culture (Louwe Kooijmanns et al, 2005; Schinkel, 2005; van dem Broeke, 2005). Some individual site abandonments have been noted elsewhere in Belgium (Krause, 2005). There is little evidence for the abandonment of farmland - indeed, there is a period of apparent expansion of production in order to provide supplies for the Roman army on the Rhine frontier. This latter case is, of course, substantially different from the other two, as the aggressor operated on a state level of organisation, which facilitated a rapid change in the control and administration of the conquered territory.

Climatic and resource problems could well explain the abandonment in Arizona. However, it is harder to understand the abandonment that seems to have occurred in the Thames Valley. The end of the Bronze Age saw climatic amelioration, surely making conditions even more appropriate for subsistence. The river highway apparently remained a constant resource, and the richness of the land is such that there would be no natural drawbacks to continued settlement. Indeed, by the Middle and Late Iron Ages, settlement and land use along the valley is widespread and dense, with large villages closely spaced both within the valley itself, and at the heads of the various tributaries. Where did the people of the Bronze Age settlements go in the period between 700 and 500 BC? One suggestion is that their new villages are to be found along the springlines above the valley (Needham, 1980) - but so far, no such settlements have come to light. Nor are they apparent on the downs immediately beyond. If the pottery dating sequences are wrong, and occupation continued in the LBA sites down to 500BC (to be abandoned at that later date), then the material culture of the people must have been very poor, leaving virtually no trace to be recovered.

J.D. Hill (2006, pers.comm.) has identified this question as one which should be addressed (among others) for this period of transition as a priority for the understanding of the later prehistoric period in southern Britain, and it is to be hoped that this will be a focus for future research.

These correlates, taken together or singly, are surely

important as indicators of the results of conflict as well as of other agents of change.

xxxi) Changes in settlement location

xxxii) Changes in settlement morphology

Beyond the changes mentioned in previous correlates, which may be variously interpreted as preparatory for defence or as resultative, long-term changes in settlement location and morphology are only really visible in the case of Gallia Belgica, as in the other two case studies we do not know where the populations went after the abandonment of their settlements. In the Thames Valley, there is little data on site morphology after the period of agglomeration; few sites have been excavated to their complete extent, and so conclusions about trends in morphology cannot be drawn.

The evidence from the Actiparc sites (Blancquaert & Prilaux, 2003), and from studies elsewhere in the region of Gallia Belgica, seem to show an organisation of sections of the population into settlements more easily controlled by the Roman authorities [paralleled, perhaps, by the movement of populations in Britain after AD43 from hilltop locations to new urbanised population centres with regular street grids (e.g. Mertens, 1983), such as Dorchester or Canterbury] or realigned along the new road networks, to facilitate the movement of produce for military needs. Most of the fortified centres become depopulated (Roymans, 1990). In the rural areas, new villa type or *ferme indigène* settlements began to appear, particularly alongside the new roads and around the new towns (Walthew, 1982). On the less fertile soils of Flanders, villa settlements are rarer, but new smaller settlements were founded along the roads and navigable waterways (Vermeulen & Bourgeois, 2000). Once again, this reflects the higher level of organization at work, the Roman state apparatus.

It is possible that these correlates can only work as indicators of the result of conflicts when this state level of change can be imposed, and thus they offer little assistance in prehistoric cases, unless the populations can be traced to new locations.

xxxiii) Changes in biodiversity due to human relocation

Apart from the obvious colonisation of abandoned areas by wild species (for which, presumably, confirmatory evidence could be supplied with a large enough programme of environmental sampling), evidence for this correlate in the case studies comes from the changes in vegetation in the Grasshopper Pueblo area mentioned above (Ezzo, 1983; Holbrook, 1982), and some indications of silting and abandonment in Gallia Belgica (Defgnée & Munaut, 1996) and around Heathrow in the Late Bronze Age (Yates, 1999).

This correlate could be a useful one, but the kind of large scale sampling strategies required have not, as yet, been instigated in any of the case study areas.

xxxiv) Changes in health profiles/standards of living

A certain degree of evidence has been recovered from Arizona populations at Grasshopper Pueblo to indicate a decline in health profiles (Reid & Whittlesey, 1999; Berry, 1985), although whether these are a result of maladaptation in the response to climatic change or to aggression is unclear. Neither of the other case studies has produced data that would be useful in this respect, although a fall in health and living standards could be surmised from historical sources in Gallia Belgica after the conquest. None of the case studies offered evidence on standards of living that are sufficiently indicative. This correlate retains potential, but can only be useful in specifically appropriate contexts where social, ritual and environmental conditions are appropriate.

xxxv) Artefactual and cultural change

The distribution of pottery and other goods has been discussed above. Other cultural changes are less obvious in the two prehistoric cases. In the very rare Thames Valley sites that show any indication of surviving into the Early Iron Age, even the presence of iron tools or weapons is almost completely absent, and there seems to be very little other artefactual change. The subsequent abandonments effectively end the archaeological record in this respect.

In Gallia Belgica, these changes are much more obvious, albeit not immediate. The introduction of Romanised goods, money, and settlement forms (Wightman, 1985) took several decades to become widespread, especially in the north of the region, because of the various exigencies related to control of populations, financial reorganisation, and Roman politics in Italy and elsewhere (van den Broeke, 2005; Hamilton, 1996). However, by about 20BC, and especially with the impetus towards 'Romanisation' given by the emperor Augustus, the region was permanently altered, the results of which are still be visible in the modern day. The founding of towns with their concomitant councils and structures, the development of *fermes indigènes* and villa estates, the building of roads and ports, the changes in agricultural and industrial production, such as the Rhine pottery industry (Drinkwater & Vertet, 1992) and even the introduction of a new language, effectively ended much of Gaulish and Belgic culture and identity in the region.

While such changes are not evident in the prehistoric case studies used in this discussion, it may be useful to reconsider other examples of apparent cultural change or imposition elsewhere with a possible interpretation relating to the outcome of aggression, and so this correlate may retain usefulness.

xxxvi) Changes in ritual and burial practices

The only evidence for change in ritual and burial practices in the Arizona case study, beyond the atypical incidences of burial which occur across the region in the PIV phase already discussed, is the construction of the Great Kiva at Grasshopper Pueblo (Reid & Whittlesey, 1997; Rice, 1989).

In the Late Bronze Age of the Thames Valley (as elsewhere in the country at that time), change is evident but not easy to understand. The barrow cemeteries of the Middle Bronze Age settlements sited along the heights above the valley were abandoned at the same time as the villages themselves. In the Late Bronze Age settlements in the valley bottom, evidence for ritual and disposal of the dead is very scarce, and those examples which exist are almost certainly atypical. The first change to note is the complete absence of burial tumuli. The amount of metalwork and other material found in the river and other watery contexts would suggest that the river played a part in funerary rites; disposal of the dead in the (east-flowing) water (whole or cremated) would seem likely, with the deposition of weapons and other artefacts as an accompanying rite. The relative absence of skeletal material from the river might suggest that cremation was practised - there are a number of 'burnt mounds' identified in several locations which could conceivably be identified as funeral places, but if cremation was the normal rite, it would be unsurprising that little evidence remains. Pyres built by the waterside would be quickly scoured away by winter floods (and tidal flows); charcoal and cremated bone would float away, and fire-heated stones would be tumbled and moved. Most of the river banks of the Thames consist of mud, which would retain little evidence of fire and that evidence is easily destroyed. What caused this change of ritual is unknown; it is clearly not localised, as a similar change is noted countrywide, and it appears to continue well into the Iron Age throughout most of Britain.

The finds that may be associated with feasting could have a ritual connection, perhaps as part of funerary or social-competitive events similar to potlatches (Barrett et al, 2001 etc.). but this cannot be confirmed.

In Gallia Belgica, change becomes evident by the Augustan period, with the appearance of Roman style cemeteries fringing the new towns, monuments to Roman soldiers and other citizens, epigraphic dedications to Roman and other deities, and iconographic depictions in various forms.

There appears to have been some continuity of ritual in Flanders (Vermeulen & Bourgeois, 2000; Lambot, 1996) and minor changes in burial ritual noted at Wederath-Belginum (Hamilton, 1996) but marked differences elsewhere (Haselgrove, 1984). Iron Age temples and sanctuaries were, in some cases, robbed by the Roman soldiers (ibid) and some have shown evidence of a Roman military presence (at La Chaussée-Tirancourt (Reddé, 1996) and Ribemont-sur-Ancre (Brunaux, 2000)). Wellington (2001) has identified sites founded after the Conquest, generally in a Roman style, which may represent a psychological form of conquest, taking over the apparatus of local beliefs.

Overall, this correlate is problematic, and may be, in the end, too fraught with complexities to be useful as an indicator for warfare in most instances.

xxxvii) Territoriality developing

The communities of Arizona seem to have developed fairly clear territorial identities around the larger pueblos, with possible conflict and/or competition between pueblos in the Tonto Basin (Reid & Whittlesey, 1997) during their periods of change, and in the Thames Valley region the emergence of more formal territories has been proposed (Lambrick, 1995) but in both cases, the almost complete abandonment of the regions at the end of their respective periods leaves us without further answers. The Thames study area contains none of the hillforts that developed in the Iron Age and there is no clear connection between the Late Bronze Age populations and the later hillfort sites to the north and south of the valley. The Roman organisation of the territories of Gallia Belgica and the formation of the *civitates* (Haselgrove, 1990a) is a matter of rather better historical and epigraphic record, but little in the way of specific archaeological evidence and once again, we are faced with a correlate that may be too obscure and difficult to be of use.

Results of the Case Studies

The evidence derived from the use of the correlates in each of the three case studies, when taken quantitatively, presents a far from useful set of results.

Overall, more correlates are met for the incidence of warfare in late prehistoric Arizona than for the other two studies (with 25% more correlates being fulfilled than for the Late Bronze Age Thames and 30% more than for Gallia Belgica).

The correlates for Arizona are strongest in the Causal and Functional groups; the Late Bronze Age Thames is most strongly indicated in the Preparatory group, and Gallia Belgica is most strongly indicated in the Resultative group. However, warfare in the Late Bronze Age Thames is least indicated in the Causal and Resultative groups, whilst in Gallia Belgica it is least indicated in the Preparatory and Functional categories.

The following table (*Table 24*) summarises these conclusions:

Causal Correlates	**Comments**	**Rating of correlate**
i) Climatic deterioration over the period	The correlate is strong, but there are problems associated with time delays within which the effects of climate change become apparent to contemporary societies, and within the archaeological record.	Strong
ii) Demographic rise	The correlate seems to have some strength; data on population numbers, however, is often lacking. This correlate is particularly strong when related to increased resource stress.	Strong
iii) Presence of other subsistence/social threats	A 'wild card' correlate that needs to be related to clearly identifiable external changes and effects.	Medium
iv) Increase in social complexity	Whilst a popularly cited correlate, this appears to be a relatively weak candidate, and may, indeed, be incidental or coincidental.	Weak
v) Territoriality developing	This correlate can equally be identified as an effect as a cause of hostilities, and is therefore a weak indicator.	Weak
vi) Changes to access for resources	A strong correlate if a pragmatic motive for warfare is assumed besides other political or social motivations	Strong
Preparatory correlates	Comments	Rating of correlates
vii) Appearance of fortifications	A weak correlate, with problems related to cause and effect, as with Correlate v above, and with the need for closer dating techniques	Weak
viii) Proliferation of weaponry	In itself, this correlate is weak, but must be included* because it speaks to the potential for hostility, and the knowledge and preparedness of communities for aggression	Medium/Weak*
ix) Iconography possibly related to warfare	A very weak correlate, probably to be excluded	Weak
x) Abandonment of settlements xi) Appearance of defensible settlement strategies xii) Changes in settlement location xiii) Changes in settlement morphology	These four inter-connected correlates possess some strength, particularly where the observed changes are seen to be maladaptive for subsistence. They are, however, subject to problems of interpretation and dating	Medium
xiv) Appearance of guard settlements	A potentially useful correlate, where sufficient of a region has been investigated	Medium/Strong
xv) Settlement locations related to line-of-sight links	A potentially useful correlate that requires much more research than is presently available for most cases.	Medium/Strong
xvi) Change in subsistence patterns	A potentially useful correlate; much might depend on the ability to identify the rate of change and the relationship between anthropogenic and climatic or other causal factors	Medium/Strong
xvii) Change in trading patterns	This correlate may be usefully indicative, but it's use depends on relatively close dating and the identification of change agents.	Medium
xviii) Unoccupied zones developing	A potentially useful correlate, depending on the fullness of survey of regions.	Medium
Functional correlates	Comments	Rating of correlates
xix) Burned sites	A potentially useful correlate, depending on survival of evidence, the nature of the burning episodes and the nature of site treatment after the event.	Strong
xx) Unburied bodies/ weapons trauma (human and animal)	A weak correlate, despite its apparent obvious quality	Weak
xxi) Weapons damage	A weak correlate unless the evidence can be specifically contextualised	Weak
Resultative correlates	Comments	Rating of correlates
xxii) Change in subsistence practice	A potentially useful correlate; much might depend on the ability to identify the dating and rate of change and the relationship between anthropogenic and climatic or other causal factors.	Medium/Strong
xxiii) Change in trading practice	On its own, this correlate is not indicative of warfare, but it may be useful when viewed in combination with other variations in commercial and productive behaviour.	Medium
xxiv) Trophy collection and memorials xxv) Iconography possibly related to warfare	Relatively weak correlates, these both require close contextualisation and a consideration of the relative strengths of whether they are indicators of actual warfare, or of political propaganda	Weak

Causal Correlates	Comments	Rating of correlate
xxvi) Deviation in sex ratio	A generally weak correlate, with inherent difficulties related to sparsely of evidence	Weak
xxvii) Depopulation xxviii) Unoccupied zones developing xxix) Population movement xxx) Abandonment of developed arable	This group of correlates is strong; they are, however, rarely addressed. There is an inherent problem for archaeologists in this respect; exploration of one or a group of related sites which display abandonment can rarely be followed up by a matching degree of research into possible candidates for relocation without specific clues and research resources.	Strong
xxxi) Changes in settlement location xxxii) Changes in settlement morphology	It is possible that these correlates can only work as indicators of the result of conflicts when a state level of change can be imposed, and thus that they offer little assistance in prehistoric cases, unless the populations can be traced to new locations.	Medium/weak
xxxiii) Changes in biodiversity due to human relocation	A potentially useful correlate that would require the instigation of programmes of large scale sampling, not yet generally available	Medium/strong
xxxiv) Changes in health profiles/standards of living	A potentially useful correlate, depending on the availability of cemetery populations and the existence of appropriate burial rites, but generally these are unavailable	Medium/weak
xxxv) Artefactual and cultural change	A relatively strong correlate where other causes of change may be discounted.	Strong
xxxvi) Changes in ritual and burial practices	A problematic correlate that may be too complex to be useful.	Weak
xxxvii) Territoriality developing	This correlate is also potentially too complex or too lacking in confirmatory evidence to be of use in most cases.	Weak

Table 24: Assessment of the Correlates

These tables may be summarised thus: *(Table 25)*

	Most present -----------------------------	-----------------------------	**Least Present**
Causal correlates	**Eastern Arizona**	**Gallia Belgica**	**MiddleThames LBA**
Preparatory correlates	**Middle Thames LBA**	**Eastern Arizona**	**Gallia Belgica**
Functional correlates	**Eastern Arizona**	**Middle Thames LBA**	**Gallia Belgica**
Resultative correlates	**Gallia Belgica**	**Eastern Arizona**	**Middle Thames LBA**

Table 25: Case studies in relative order of presence of correlates

Causal correlates

Strongest	Eastern Arizona
Medium	Middle Thames LBA
Weakest	Gallia Belgica

Preparatory correlates

Strongest	Eastern Arizona
Medium	Middle Thames LBA
Weakest	Gallia Belgica

Functional correlates

Strongest	Eastern Arizona
Medium	Middle Thames LBA
Weakest	Gallia Belgica

Resultative correlates

Strongest	Gallia Belgica
Medium	Middle Thames LBA
Weakest	Eastern Arizona

Table 26: Correlate results in order of strength

Clearly these results are related to the quality and quantity of evidence available, and to the value which may be placed on the strength or weakness of each correlate. If the correlates are accorded relatively arbitrary values for strength, based on the discussion above and the evidence from these specific case studies, the following pattern can be discerned.*(Table 26):*

Given the historical confirmation of warfare in Gallia Belgica, and the special circumstances of cause and preparation, it seems necessary to conclude that the resultative correlates at least offer some confirmation of the historical evidence, but to note the general absence of functional correlates. Given that these are the very correlates most often cited for the recognition of warfare in prehistory, it is salutary to observe that they are not visible in this, the most definite case of actual warfare. Clearly, this has major implications for the recognition of warfare in earlier periods.

The case of Eastern Arizona offers a stronger overall impression that warfare may have taken place; the case study is weakest in the resultative correlates, but this result is generally related to lack of data rather than negative evidence. It would seem, therefore, that there is some corroboration of the bias in the literature of this area towards the occurrence of warfare in the period considered.

As might be expected, the Middle Thames Late Bronze Age demonstrates far less confirmatory evidence in all groups of correlates overall, although not by a large margin. It might, therefore, be fair to say that warfare in the period remains a possibility, but that, in fact, much of the strongest evidence relates to preparedness for aggression, and social and other changes related to this preparation, than to actual conflict. It may well be that the preparations made for defence by these communities were, in the event, successful in averting war.

Discussion of the correlates

There are a number of questions which have arisen as a result of these case studies. In the first place, can correlates of the type proposed here help in the recognition and confirmation of periods of war in prehistory? Are the correlates chosen the right ones? Are the groupings of correlates helpful or not? Can one set of general correlates successfully cope with the number and form of variables in different cases, places and times, or must each occasion be treated separately? To what extent does the formation of a set of correlates lend itself to the creation of a self-fulfilling prophecy? How are alternative interpretations of evidence to be assessed and given value?

To regard warfare as a process is not to limit the search for evidence of its occurrence to the kind of interpretative approach that is dominated by systems theory and positivism; it is rather an attempt to recognise warfare as the type of human event which can extend itself over a long time period and a wider set of behaviours than has been considered in some archaeological discussions of the topic. The proposal of correlates may provide a way to create a methodological context for the recognition of the inter-relatedness of these behaviours; it does not offer any restriction of the possibilities for a variety of explanations for them.

The correlates may offer a basis for the reassessment and negotiation of relative interpretations within a more developed concept of war, its stages and symptoms. Shanks & Hodder (1995:15) maintain that a focus on context permits a sensitivity to diversity, and also that most contexts are part of a larger group of contexts, all in some way capable of affecting each other. The more contexts and evidence that can be brought together, "the more likely is one able to make statements about meaning" (ibid).

It is important, therefore, to approach topics such as warfare not only in ways which seek to recognise as many of their associated behavioural contexts as possible, but also to recognise the need to initiate studies on a sufficiently large geographic scale, and to attempt to create a homogeneity in the gathering of evidence that will allow meaningful comparison.

It is abundantly clear that one of the major problems in this form of study (as indeed in most archaeological investigations) is a paucity of sufficient, comparable evidence. Even in a single region, sites may have been investigated at different times, under varying conditions, with unequal budgets and significant differences in the level of technological analysis available to aid interpretations, to say nothing of the application of different theoretical mind sets in the design and analysis of the results. Another urgent need is for a better understanding of the chronological relationship of sites, their periods of use, contemporaneity and date of abandonment.

Having said this, it is necessary to consider what *can* be deduced. The causal correlates proposed do, on the whole, offer some direction. Climatic and demographic change do seem to be implicated in major social, subsistence and settlement changes. Changes in social complexity, access to resources and territoriality are complex areas to study; they *may* be implicated in causes for warfare, they may result from warfare, or they may be unconnected to aggression. Close dating of observed changes is often not possible, and their interpretation must depend on a wide range of other evidence. It is possible that the use of these correlates might be most efficacious in the negative - to indicate the absence, rather than the presence, of war. Thus, if the changes are observed, but none (or very few) of the preparatory and functional correlates are present, then warfare probably did not take place. However, the evidence in the Gallia Belgica case study seems to demonstrate that this would be a dangerous assumption.

A number of the preparatory correlates (which are subject to the same problems of dating and availability of evidence) may well represent attempts to avoid conflict - fortifications, settlement movements, proliferation of weaponry, development of guard settlements and line-of-sight links may all be ways to make sure attacks do not take place, a form of Cold War investment. This certainly may have the implication that warfare is a possibility, but not that it actually occurred. Changes in subsistence and trade patterns may have multiple causes. It is very difficult to assign these to a specific set of circumstances or to a limited time period.

Where change is maladaptive, however, reasons must be sought, and here aggression, or its threat, may well be implicated.

It may well be believed that the functional correlates should be the most straightforward to consider. However, the case studies amply demonstrate that this is not the case. Each correlate, even where present, is subject to a multiplicity of possible explanations.

Changes that result from periods of warfare have rarely been investigated in any detail, yet certainly in the Gallia Belgica case study, these were the most visible correlates found.

Resultative changes in subsistence practice and trading practice are most clear in the Gallia Belgica case in ways that might be directly attributable to the after effects of war; changes in the Arizona cases are less clear, and may be attributed to climatic change, although their apparent maladaptation may be an indicative factor. No clear evidence can be derived from the Late Bronze Age Thames study, due to a lack of close dating and the abandonment of the sites.

The correlates relating to trophy collections, memorials and iconography gave no assistance in the case studies, with the exception of the Roman examples. Useful examples of these features are likely to be rare, and therefore the usefulness of these correlates is likely to be extremely limited.

Few cases of possible prehistoric war are likely to have sufficient burial population evidence to confirm deviation in sex ratios, and even if large numbers of burials are found, cultural and social customs may well be responsible for apparent gender inequalities. The same applies to data on health and standards of living.

Depopulation and abandonment are, perhaps, more useful - but the greatest use could be made of data associated with these correlates if it is possible to trace the eventual destination of the moving populations. In neither the Arizona nor the Thames case studies was this possible. The same parameters apply to the correlates associated with settlement location and morphology and changes in biodiversity.

Lack of data also plagues the discussion of artefactual, cultural and ritual change; once again, the best evidence comes from Gallia Belgica, and here it is argued that the (albeit slow) progress of domination of Roman-style artefacts, architecture and burial customs is great enough to demand explanation beyond the assumption of peaceful cultural change. These correlates seem to be stronger indicators of great enough change to require mega-explanations, of which warfare may be an example. Territoriality development is also relatively well-represented, although this is another occasion where the problems of cause and effect, or of preventative strategies, need to be addressed.

In summary, the apparently strongest correlates can be characterised as those concerned with environmental and demographic change with concomitant changes in resource acquisition, depopulation and abandonment (and evidence of resulting biodiversity changes), morphological and locational changes to settlement patterns, and artefactual and cultural changes. Only the correlate relate to burning of sites remains strongly indicated in the functional category. Most of the indicators commonly cited for the recognition of warfare, therefore, are in fact apparently amongst the weakest correlates.

However, the stronger correlates are all, in themselves, capable of a multiplicity of alternative explanations and interpretations. The challenge will be to create alternative hypotheses for their occurrence, and through analysis, to reject such hypotheses that do not offer a similar set of confirmatory presences to that of warfare.

The creation of a list of correlates cannot confirm that warfare took place, but it is at least a place to start, and a foundation from which alternative hypotheses can be assessed.

Implications for future research

Within the framework provided by a suitable set of correlates, reassessment of existing evidence should be accompanied, where possible, by a greater emphasis on the collection of data that will provide more evidence of climatic and environmental conditions, from as great a number of samples as can be gathered, to elucidate the rates of change affecting past communities, and to help assess the forms and efficacy of mitigation and adaptation strategies that may have been employed in settlement and subsistence choices. Changes in behaviour that might seem to be maladaptive need, then, to be investigated thoroughly, as should evidence of settlement or subsistence abandonment. Most archaeological sites are, *de facto,* abandoned. It is not sufficient to simply note the end of occupation around a particular date, without an effort to understand the causes of cessation of activity, and an attempt to find out what happened to the inhabitants subsequently. If abandonment is apparently visible over a reasonably large geographic area, and within a reasonably defined time scale, any vague assumptions about settlement or subsistence shift, or ritual destruction, simply will not do. Programmes should be designed to find those shifted settlements, or to explain those changes in subsistence practice; we need to be able to search for what happened next. It may rarely be possible to investigate the whole of a site or a region in any detail. Comparative studies, therefore, must be particularly useful, bringing together evidence from a range of sites and contexts. This presupposes that new investigations can be designed to gather and publish data in a manner which makes comparison straightforward. It would also be useful to collect, study and archive samples for environmental analysis on a greater scale than is often the case, although financial and accommodation constraints probably make this a difficult goal to achieve in the current state of archaeological provision.

Discussion about, and recognition of, appropriate groups of correlates for a number of types of behaviour could assist in the construction of vastly improved and more subtle methodological contexts, in which the search for meaning and understanding could be developed on an extended and focussed evidential base. Testing of correlate groups might help to refine ideas about the ways in which societies could have reacted in certain situations, warfare among them, and thus contribute towards the creation of a potentially useful tool for the informing of future research designs.

The development of correlate groups demands an analytical approach to the understanding of the relationships between types of observable behaviour recognised in archaeological evidence, and could help researchers to avoid the dangers of assumptions and expectations dependent on the political and behavioural present.

Appendix A

Plants indicative of abandonment/waste (Britain)

Aegopodium podgravia	ground elder
Anthemis cotula	stinking chamomile
Anthriscus caucalis	bur chervil
Aramanthus retroflexus	coomon aramanth
Arctium lappa	great burdock
Arctium mirus	lesser burdock
Aristolochia clematitis	birthwort
Artemisia absinthium	wormwood
Artemisia vulgaris	mugwort
Atriplex petula	common orache
Ballota nigra	black horehound
Brassica nigra	black mustard
Bryonia cretica	white bryony
Calystegia sepium	hedge bindweed
Calystegia silvatica	great bindweed
Camelina sativa	gold-of-pleasure
Cannabis sativa	hemp
Carduus acanthoides	welted thistle
Carduus nutans	musk thistle
Carduus tenuiflorus	slender thistle
Chamomilla recutita	scented mayweed
Chenopodium bonus-henricus	Good King Henry
Cichorium intybus	chicory
Cirsium arvense	creeping thistle
Cirsium vulgare	spear thistle
Cicuta virosa	cowbane
Conium maculatum	hemlock
Conopodium majus	pignut
Convolvulus arvensis	field bindweed
Coronopus squamatus	swine cress
Crepis capillaris	smooth hawksbeard
Crepis foetida	stinking hawksbeard
Cuscuta europaea	greater dodder
Descurainia sophia	flixweed
Epilobium angustifolium	rosebay willowherb
Erysimum cheiranthroides	treacle mustard
Euphorbia platyphillos	broad-leaved spurge
Fallopia convolvulus	black bindweed
Foeniculum vulgare	fennel
Gallium aparine	cleavers
Heracleum sphondylium	hogweed
Hypericum perforatum	perforate St. John's wort
Lactuca serriola	prickly lettuce
Lamium album	white dead-nettle
Lamium purpureum	red dead-nettle
Lavatera cretica	small tree-mallow
Linaria vulgaris	common toadflax
Malva sylvestris	common mallow

Marrobium vulgare	white horehound
Medicago lupulina	black medick
Nuphar pumila	least yellow water-lily
Parietaria officinalis	pellitory
Petroselinum crispum	garden parsley
Polygonum aviculare	common knotgrass (and other knotgrasses)
Polygonum persicaria	pale persicaria
Potentilla reptans	creeping cinquefoil
Ranunculus sardous	hairy buttercup
Rorippa sylvestris	creeping yellow-cress
Rosa arvensis	field rose
Rubus fruticosus agg.	bramble
Rumex obtusifolius	broad-leaved dock
Senecia jaconaea	common ragwort
Senecia vulgaris	groundsel
Solanum dulcamera	bittersweet
Solanum nigrum	black nightshade
Sonctus asper	prickly sow-thistle
Stellaria media	common chickweed
Tanacetum vulgare	tansy
Tordylium maximum	hartwort
Torilis nodosa	knotted hedge parsley
Tragopogon pratensis	goatsbeard
Tusilago farara	coltsfoot
Urtica dioica	common nettle
Urtica urens	small nettle
Verbascum thepsus	great mullein
Verbascum virgatum	twiggy mullein
Verbena officinalis	vervain

Note:

While most of these plants may well be present in small waste areas, their increase may be indicative of abandonment processes; a fall in numbers of common weeds of cultivation and of plants associated with anthropogenic activity (plants of trampled ground etc.) will tend towards the same indication. Only native species and ancient introductions have been included in this list. [Compiled from various sources, particularly L. Grundy (pers.comm.)].

Appendix B

Perry Mesa - Sites of the Pueblo III Period*

Upper Black Mesa group				
Alkali Canyon Ruin	19+11 rooms	elevation (ft)	3,320	
Quadruped Site	12+6		3,320	
Running Deer Site 2	20-30		3,060	
Little Rattler Site	30-50		3,060	
Site NA No. 13,311	9-12+		3,440	
Richinbar Site	6-8		3,450	
Kayenta	?		3,490	
Site NA No. 13,322	2-5		3,420	
Site NA No. 13,308	2		3,560	'fort'
Middle Black Mesa group				
Site NA No. 13,294	1-2		3,390	
Site NA No. 13,320	2		3,410	
Site NA No. 13,291	15		3,390	
Site NA No. 13,292	16		3,340	
Site NA No. 13,342	1-3		3,350	
Site NA No. 13,340	10		3,330	
Above Black Canyon group				
Henrie Site	c22 (775 sq.m.)		2,885	hilltop habitation
Euler Site	20			hilltop habitation
Spanish Hill	2			retreat
Water Tank	0			retreat
Horsethief Basin group				
South Fort (Dunbar)				hilltop fort
East Fort (Galatti)	(650 sq.m.)			hilltop fort
Dugas area				
Estler Peak	4			hilltop fort
Cleater/Cordes area				
Golden Turkey Pueblo	c12		3,000	pueblo
Site NA No. 5847	9			defensible
Dripping Springs				hilltop habitation
DeNoyelles Site	16 (743 sq.m.)			hilltop habitation + plaza
Mesa Site	?			fort/lookout
Townsend Butte				fort/lookout
Mayer Area				
Myers Fortress	5			fort/lookout
Joanne Cline Site	(88 sq.m.)		4,000	fort/lookout

Earl Cline Site	14+	4,260	hilltop habitation
Copper Mountain North	3	4,840	fort/lookout
Copper Mountain South	(175 sq.m.)	5,026	fort/lookout
Jagged Tooth Ruin	23-30		hilltop habitation
Nash Site	c15	5,142	hilltop habitation
Triangle M Site	c10		pueblo
Fence Tank Fort			fort/lookout
Site NA No. 4636	5+		pueblo
Humboldt/Poland Junction area			
Lower Stoddard Ruin	(743 sq..m.)		pueblo
Site NA No. 2411	20+		pueblo
Site NA No. 4364	70		pueblo
Humboldt Ruin	18		pueblo
Woolsey Ranch Pueblo	80+ (2,704 sq.m.)		pueblo
Site NA No. 4365	19		'fortified'
O'Brien Ruin	'small'		pueblo
Site NA No. 4353	c10		pueblo
Site NA No. 4354	(9,750 sq.m.)		four small pueblos
Site NA No. 4361	c7		pueblo
Site NA No. 4360	c19		two pueblos
Boulder Maze	5		fortified pueblo
Fitzmaurice Ruin	25+	5,100	fortified pueblo (burned)
Agua Fria - Lake Pleasant North			
Site T:3:1	? rooms		walled hilltop habitation
Indian Mesa	17-22	2,340	hilltop habitation
Boulder Creek Ruin	(1,793 sq.m.)		hilltop retreat
Fort Tule	(82 sq.m.)	2,330	fort/lookout
Agua Fria Ruin	1 (c3,000 sq.m.)		religious enclosure?
Secombe Site		2,405	lookout
Sites below Lake Pleasant and Cave Creek Dam			
Site T:7:34	?		fort/hilltop habitation
Fort Mountain	(2,500 sq.m.)		hilltop retreat
Shaw Butte	(586 sq.m.)		hilltop retreat
Daisy Mountain/Cave Creek System			
Site T:4:8	87 (4,212 sq.m.)		hilltop centre
Site T:4:3	(1,080 sq.m.)		hilltop habitation
Site T:4:6	(7,412 sq.m.)		hilltop habitation
Site T:4:7	(531 sq.m.)		fort
Site U:1:7	30 (4,050 sq.m.)		hilltop habitation
Site U:1:13	40 (4,048 sq.m.)		hilltop habitation
Site U:1:18	(312 sq.m.)		hilltop habitation
Site T:4:5	(1,092 sq.m.)		hilltop retreat
Site U:1:21			hilltop retreat
Site U:1:10	(465 sq.m.)		hilltop retreat
Site U:1:20	(3,000 sq.m.)		hilltop retreat
Site U:2:27	(2,560 sq.m.)		hilltop retreat
Camp Creek System			
Sears-Jay Site	c40 (8,097 sq.m.)		hilltop habitation

Blue Mountain	15 (3,000 sq.m.)	hilltop habitation
Site U:1:31		habitation destroyed by fire
Site U:1:27		fort
Kentuck Mountain Fort		fort
Lower Verde Hilltop System		
Brazaletes Site	100+ (4,849 sq.m.)	hilltop centre
Site U:2:2	20-35	hilltop habitation
St Clair Peak Fort		fort/lookout
The Citadel		hilltop habitation
Site O:14:152	c25	?compound
Tangle Fort	(416 sq.m.)	fort/lookout

* Variations in data reflect type and quality of recording; data recovered from mixed sources including aerial survey, ground survey, excavation and antiquarian records *(after Wilcox, Robertson & Wood, 2001).*

Bibliography

Adler, M.A., van Pool, T., & Leonard, R.D., 1996 'Ancestral Pueblo Population

Aggregation and Abandonment in the North American Southwest', *Journal of World Prehistory,* **10**(3):375-438

Agache, R., 1979 'Nouveaux apports des prospections aériennes en archéologie pré-romaine et romaine de la Picardie' *Cahiers Archéologique de la Picardie* **6**, 33- 90

Allen, I.M., Britton, D. & Coghlan, H.H., 1970 *Metallurgical Reports on British and Irish Bronze Age Implements and Weapons in the Pitt Rivers Museum,* Oxford: Oxford University Press Occasional Papers on Technology, 10

Amanpour, C., 1996 'Balkans war leaves devastation behind' http://www.cnn.com accessed 22/05/03

Arnold, C.J., 1995 'The Archaeology of Inter-Personal Violence' *Scottish Archaeological Review,* **9** & **10**:71-79

Austin, K., 2000 'The Mountain Patayan People of West Central Arizona' in Motsinger, T., Mitchell, D.R. & McKie, J.M. (eds) *Archaeology in West Central Arizona:Proceedings of the 1996 AAC Prescott Conference,* Prescott: Sharlot Hall Museum Press, 63-74

Avery, M., 1986 " 'Stoning and Fire' at hillfort entrances of southern Britain" *World Archaeolog* **18**(2): 216-230

Bahn, P., 1998 *The Cambridge Illustrated History of Prehistoric Art,* Cambridge: Cambridge University Press

Bahn, P.G., (ed) 1996 *Tombs, Graves and Mummies,* London: Weidenfeld & Nicolson

Baillie, M.G.L., 1998 'Hints that Cometary Debris Played some Role in Several Tree-Ring Dated Environmental Downturns in the Bronze Age', in Peiser BJ, Palmer T & Bailey ME (eds) 1998 *Natural Catastrophes During Bronze Age Civilizations : Archaeological, geological, astronomical and cultural perspectives,* Oxford: BAR Int Ser 728, 109-116

Balikci, A., 1970 *The Netsilik Eskimo,* Garden City (NY): Natural History Press Barber, K.E., 1982 Peat-bog stratigraphy as a proxy climate record' in Harding, A. (ed) *Climate Change in Later Prehistory*, Edinburgh UP, 103-113

Barrett, J., 1980 'The evolution of Later Bronze Age Settlement' in J. Barret & R. Bradley (eds), *Setttlement and Society in the Later British Bronze Age,* Oxford, BAR 83 (I), 77-100

Barrett, J.C., Lewis, J.S.C. & Welsh, K., 2001 Perry Oaks a history of inhabitation, part 2, *London Archaeologist* **9**(8): 221-227

Barth, F., (ed) 1969 *Ethnic groups and boundaries: the social organisation of culture difference,* London: Allen & Unwin

Bayard, D., 1996 'La Romanisation des campagnes en Picardie à la lumière des fouilles récentes: problèmes d'échelles et de critères', *Revue archéologique de Picardie, no. sp.* ***11,*** 157-184

Beltran, A., 1982 (tr. M. Brown) *Rock Art of the Spanish Levant,* Cambridge: Cambridge University Press

Bénard, J., 1996 'Alésia: du mythe à l'archéologie' in Reddé, M. (ed), *L'Armée Romaine en Gaule,* Paris: Edition Errance, 40-65

Berry, D.R., 1985 'Aspects of Paleodemography at Grasshopper Pueblo, Arizona" in Merbs, C.F. & Miller, R.J., (eds) *Health and Disease in the Prehistoric Southwest,* Tempe: Arizona State University Anthropological Research Papers No. 34 , 43-67

Bigelow, R., 1975 'The role of competition and co-operation in human evolution' in Nettleship, M.A., Dale Givens, R., & Nettleship A., (eds) *War, Its Causes and Correlates,* The Hague and Paris: Mouton, 235-267

Billman, B.R., 1997 'Population pressure and the origins of warfare in the Moche Valley, Peru' in Paine, R.R., (ed) *Integrating Archaeological Demography - Multi-disciplinary approaches to prehistoric populations,* Carbondale: Center for Archaeological Investigations, Southern Illinois, Occasional Reports No. 24, 285- 310

Blagg, T. & Millett, M. (eds)., 2002 *The Early Roman Empire in the West,* Oxford: Oxbow

Blancquaert, G., Prilaux, G. et al, 2003 'Exceptionnels Vestiges de la Conquête Romaine', *Archéologia* **404,** 12-25

Bloemers, J.H.F., 2002 'Lower Germany: *plura consilio quam vi* - proto-urban settlement developments and the integration of native society' in Blagg, T. & Millett, M., (eds) 2002 *The Early Roman Empire in the West,*Oxford: Oxbow, 72-86

Bonenfant, P-P., 1998 'Les fortifications de l'Age du Fer en Hainault', in Bayard, D. (ed), *Les Celtes: Rites funéraires en Gaule du Nord entre le Vie et le 1er siècle avant Jesus-Christ: Recherches récentes en Wallonie,* Tournai: Division du Patrimoine et Maison de la Culture de Tournai, 143-148

Bourgeois, I., Cherretté, B., & Bourgeois, J., 2003 Bronze Age and Iron Age Settlements in Belgium: an overview, in Bourgeois, J., Bourgeois, I. & Cherretté, B., (eds), *Bronze Age and Iron Age Communities in North-Western Europe,* Brussel: Koningluke Vlaamse Academie van Belgie voor Wetenschappen en Kunsten, 175-190

Bourgeois, J., Bourgeois, I. & Cherretté, B., (eds), 2003 *Bronze Age and Iron Age Communities in North- Western Europe,* Brussel: Koningluke Vlaamse Academie van Belgie voor Wetenschappen en Kunsten

Bowden, M. & McOmish, D., 1989 'Little Boxes: More about hillforts' *Scottish Archaeological Review,* **6**: 12-16

Bowden, M. & McOmish, D., 1987 'The required barrier' *Scottish Archaeological Review* **4:** 97-99

Bowen, H.C., 1978 ' 'Celtic' fields and 'ranch' boundaries in Wessex', in *The effect of man on the landscape: the Lowland Zone,* eds. Limbrey S. & Evans J.G., CBA Research Report No. 21, 115-123

Boyle, A., 1992 An Analysis of the Human Remains, in Moore, J. & Jennings, D. *Reading Business Park. A Bronze Age Landscape,* Oxford: Oxford University Committee for Archaeology, p92

Bradley, R., 1998, 'Daggers Drawn: depictions of Bronze Age weapons in Atlantic Europe' in Chippindale, C. & Taçon, P.S.C., *The Archaeology of Rock Art,* Cambridge: Cambridge University Press, 130-145

Bradley, R., 1990 *The Passage of Arms,* Cambridge: Cambridge University Press

Bradley, R. 1986 The Bronze Age in the Oxford Area – Its Local and Regional Significance in Briggs, G., Cook, J., and Rowley, T., *The Archaeology of the Oxford Region,* Oxford, Oxford University Department for External Studies, 38-48

Bradley, R., 1980 Subsistence, Exchange and Technology a Social Framework for the Bronze Age in Southern England c. 1400-700 bc, in Barrett, J. & Bradley, R., (eds) *The British Later Bronze Age,* Oxford: British Archaeological Reports British Series 83 (i), 57-75

Bradley, R., 1978, *The Prehistoric Settlement of Britain,* London: Routledge & Kegan Paul

Brailsford, H N, 1906 *Macedonia: Its Races and Their Future,* London: private publication, pp36-7

Briard, J., 1997 (tr. J-P. Perre) *The Megaliths of Brittany,* Luçon: Editions J-P. Gisserot

Briard, J., 1978 edn (tr. M. Turton), *The Bronze Age in Barbarian Europe,* London: BCA

Brickley, M., 2000 'Human Skeletal Remains: putting the humans back into human ecodynamics' in Bailey, G., Charles, R. & Winder, N., (eds) *Human Ecodynamics (Proceedings of the Association for Environmental Archaeology Conference 1998)* Oxford: Oxbow, 141-146

Bridges, P.S., 1996 'Warfare and Mortality at Koger's Island, Alabama' *International Journal of Osteoarchaeology,* **6**:66-75

Bridgford, S.D., 1997aThe first weapons devised only for war, *British Archaeology,* March, 22:7

Bridgford, S., 1993 *The Swords of the Late Bronze Age in Ireland,* unpublished MA Thesis, Sheffield University

Brooks, R.L., 1993 'Household abandonment among sedentary Plains societies: behavioural sequences and consequences in the interpretation of the archaeological record' in Cameron, C.M. & Tomka, S.A., (Eds) *Abandonment of Settlements and Regions: Ethnoarchaeological and archaeological approaches,* Cambridge:CUP, 178-187

Brossler, A., Early, R. & Allen, C., 2004 *Green Park (Reading Business Park) Phase 2 Excavations 1995 – Neolithic and Bronze Age Sites,* Oxford: Oxford Archaeology Thames Valley Landscape Monograph No. 19

Brothwell, D., 1999 'Biosocial and Bio-archaeological Aspects of Conflict and Warfare' in Carman, J. & Harding, A. (eds) *Ancient Warfare,* Stroud: Sutton, 25-38

Brown, A.G. & Barber, K.E., 1985 Late Holocene paleoecology and sediment history of a small lowland catchment in central England, *Quaternary Research* **24**, 87-102

Brunaux, J.-L., 2000 'Ribemont-sur-Ancre (Somme)', *Gallia* **56** (1999: 177-283), Paris, CRNS Editions

Brunaux, J-L., 1999a 'Gallic Blood Rites', *Archaeology,* (May/June), 54-57

Brunaux, J-L., 1999b 'Ribemont-sur-Ancre (Somme). Bilau préliminaire et nouvelles hypothèses', *Gallia* **56**, 177-283

Burgess, C., 1985 Population, climate and upland settlement in Spratt, D. & Burgess, C. (eds) *Upland settlement in Britain. The second Millennium BC and after,* BAR Br Ser 143, 195-230

Burgess, C.B. & Colquhoun, I., 1988 'The Swords of Britain' *Prähistorische Bronzefunde* 4(5), Munich

Butterworth, C.A. & Lobb, S.J., 1992 *Excavations in the Burghfield Area, Berkshire : developments in the Bronze Age and Saxon landscapes,* Salisbury: Trust for Wessex Archaeology , 1992.

Caesar, G.J. (tr. S.A. Handford, 1951) *The Conquest of Gaul,* Harmondsworth: Penguin

Cahen, D., Caspar, J-P., Gosselin, F. & Hauzeur, A., 1987 'Le Village Rubané Fortifié de Darion (Province de Liège)' *Archäologisches Korrespondenzblatt* **17**: 59- 68

Campbell, C., 1986 'Images of war: a problem in San rock art research' *World Archaeology* **18**(2): 255-268

Carman, J. & Harding, A. (eds) 1999 *Ancient Warfare,* Stroud: Sutton

Carneiro, R.L., 1992 'The role of warfare in political evolution: Past results and future projections' in Ausenda,

G., (ed) *Effects of War on Society,* San Marino: Centre for Interdisciplinary Research on Social Stress, 87-102

Carneiro, R.L., 1978 'Political expansion as an expression of the principle of competitive exclusion' in Cohen, R. & Service, E.R., (eds) *Origins of the State: the Anthropologyof Political Evolution,* Philadelphia: Institute for the Study of Human Issues, 205-223

Carneiro, R.L., 1970 'A Theory of the Origin of the State', *Science* **169**: 733-738

Carroll, M., 2001 *Romans, Celts and Germans: The German Provinces of Rome,* Stroud: Tempus

Chagnon, N.A., 1983 *Yanomamö: The Fierce People,* 3rd edn., New York: Holt, Rinehart & Winston

Chagnon, N.A., 1968 'Yanomamö warfare' in Chagnon, N.A. (ed), *Yanomamö: The Fierce People,* New York: Holt, Rinehart & Winston, 170-190

Champion, T., 1982 'Fortification, ranking and subsistence' in Renfrew, C. & Shennan, S. (eds) *Ranking, Resource and Exchange,* Cambridge: Cambridge University Press, 61-66

Champion, T., Gamble, C., Shennan, S. & Whittle, A., 1992 *Prehistoric Europe,* London;Longman

Chapman, J., 1999 'The origins of warfare in the prehistory of Central and Eastern Europe' in J. Carman & A. Harding, (eds), *Ancient Warfare: Archaeological Perspectives,* Stroud: Sutton, 101-142.

Chatters, J.C., 1989 'Pacifism and the Organisation of Conflict on the plateau of Northwestern America' in Tkaczuk, D.C.& Vivian, B.C. (eds) *Cultures in Conflict: current archaeological perspectives, Proceedings of the Twentieth Annual (Chacmool) Conference of the Archaeological Association of the University of Calgary,* Calgary: University of Calgary, 241-252

Ciolek-Torello, R., 1985 'A Typology of Room Function at Grasshopper Pueblo, Arizona' *Journal of Field Archaeology,* **12**(i): 41-63

Claessen, H.J.M., 2000 *War and state formation - is there a connection?* Paper presented at the War and Society seminars, Aarhus University, 5th October 2000

Coles, J. M., 1962 'European Bronze Age shields' *Proceedings of the Prehistoric Society,* **28**: 156-190

Coles, J.M., Heal, S.V.E., & Orme, B.J., 1978 'The Use and Character of Wood in Prehistoric Britain and Ireland' *Proceedings of the Prehistoric Society,* **44**: 1- 45

Collis, J., 1981, 'A theoretical study of hill-forts' in *Hill-Fort Studies: Essays for A.H.A. Hogg, ,* Leicester: Leicester University Press, 66-76

Colquhoun, I.A., 1979 'The Late Bronze Age hoard from Blackmoor, Hampshire' in C.B. Burgess & D. Coombs, *Bronze Age Hoards,*Oxford: British Archaeological Reports No. 67, 99-115

Constandse-Westermann, T.S. & Newell, R.R., 1984 'Mesolithic trauma: Demographical and chronological trends in Western Europe' *Proceedings of the 4th European Paleopathology Conference,* Middelburg/ Antwerpen: Universitat Middelburg, 1982

Coombs, D., 1975, 'Bronze Age weapon hoards in Britain' *Archaeologica Atlantica* 1(1): 49-81

Cordell, L., 2000 'Aftermath of Chaos in the Pueblo Southwest' in Bawden, G. & Reycraft, R.M. (eds) *Environmental Disaster and the Archaeology of Human Response,* Albuquerque (University of New Mexico) Maxwell Museum of Anthropology Anthropological Papers No. 7, 179-194

Corning, P.A., 1975 'An evolutionary paradigm for the study of human aggression' in Nettleship, M.A., Dale Givens, R., & Nettleship A., (eds) *War, Its Causes and Correlates,* The Hague and Paris: Mouton, 359-387

Cotter, M., 1999 *Rivers of Blood, Rivers of Gold: Europe's Conflict with Tribal Peoples,*London: Pimlico

Cotton, J., 1991 Prehistory in Greater London *Current Archaeology* **XI**(124):No.4, 151-154

Coulston, J., 2001 'The archaeology of Roman conflict' in Freeman, P.W.M. & Pollard, A.(eds) *Fields of Conflict: Progress and Prospect in Battlefield Archaeology, Proceedings of a conference held in the Department of Archaeology, University of Glasgow,* Oxford: British Archaeological Reports 958: 23-49

Crotty, H.K., 2001 'Shields, Shield Bearers and Warfare Imagery in Anasazi Art, 1200-1500' in Rice, G.E. & LeBlanc, S.A. (eds) *Deadly Landscapes: Case Studies in Prehistoric Southwestern Warfare,* Salt Lake City: University of Utah Press, 65-84

Crotty, H.K., 2001 'Shields, Shield Bearers and Warfare Imagery in Anasazi Art, 1200-1500' in Rice, G.E. & LeBlanc, S.A. (eds) *Deadly Landscapes: Case Studies in Prehistoric Southwestern Warfare,* Salt Lake City: University of Utah Press, 65-84

Crouch, K.R. & Shanks, S.A., (eds) 1984 *Excavations in Staines, 1975-76 : the Friends' burial ground site,* London/Guildford; London & Middlesex Archaeological Society/Surrey Archaeological Society

Cunliffe, B., 2004 'Wessex Cowboys?' *Oxford Journal of Archaeology,* **28**(1), 61-81

Cunliffe, B., 1993 *English Heritage Book of Danebury,* London: B T Batsford/English Heritage

Cunliffe, B., 1984 *Danebury: an Iron Age hill-fort in Hampshire, 2vols.,* London: Council for British Archaeology

Cunliffe, B.W., 1976, *Iron Age Sites in Central Southern England,* CBA Research Report 16, London

Cunnington, M.E., 1912, 'Knap Hill Camp' *Wiltshire Archaeological Magazine* **XXXVII**, 42-65

Curwen, E.C., 1930 'Neolithic Camps' *Antiquity* **4** (13): 22-54

Dark, P., 2006 'Climatic deterioration and land-use change in the first millennium BC; perspectives from the British palynological record' *Journal of Archaeological Science* **33**: 1381-1395

Dean, R.M., 2001 'Social Change and Hunting during the Pueblo III to Pueblo IV Transition' *Journal of Field Archaeology,* **28**(3/4), 271-285

Debord, J., 1995a 'A Propos de la Chronologie de Pommiers et de Villeneuve-Saint-Germain (Aisne)', *Revue archéologique de Picardie,* **1 / 2**, 205-206

Debord, J., 1995b 'La Fortification de l'Oppidum

de Villeneuve-Saint-Germain (Aisne)', *Revue archéologique de Picardie*, **1 / 2,** 187-203

Defgnée, A. & Munaut, A.V., 1996 'Evolution de l'environment végétal du Nord de la Gaule de La Tene à l'époque gallo-romaine' *Revue archéologique de Picardie, no. sp. **11**,* 325-332

Dent, J.S. 1983 'Weapons, Wounds and War in the Iron Age', *Archaeological Journal* **140**, 120-128

Derevenski, J.S., 2000 'Material culture shock: Confronting expectations in the material culture of children' in J.S. Deverenski, (ed), *Children and Material Culture,* London & New York: Routledge, 3-16

Deyber, A 1996 'L'Après-Guerre: Bilan d'un désastre' in Reddé, M. (ed), *L'Armée Romaine en Gaule,* Paris, Editions Errance, 83-92

Dixon, P., 1989 'Crickley Hill' in C. Burgess *et al* (eds.) *Enclosures and Defences in the Neolithic of Western Europe,* Oxford: British Archaeological Reports International Series 403: 75-87

Doelle, W.H., 2000 'Tonto Basin Demography in a Regional Perspective' in Dean, J.S. (ed), *Salado*, Albuquerque: Amerind Foundation, 81-105

Drewett, P., 1977 'Excavation of a Neolithic Causewayed Enclosure on Offham Hill, East Sussex, 1976' *Proceedings of the Prehistoric Society* **43**: 201-241

Drewett, P., Rudling, D. & Gardner, M., 1988 *The South-East to AD 1000,* London: Longman

Drinkwater, J. & Vertet, H., 1992 '"Opportunity" or "Opposition" in Roman Gaul?', in Wood, M. & Queiroga, F., (eds) *Current Research on the Romanization of the Western Provinces,* Oxford: British Archaeological Reports International Series S575, 25-28

Dynes, M., 'Farmers in famine-hit Malawi resort to mob justice', *The Times,* 15th July 2002, 12

Ember, C.R. & Ember, M., 1997 'Violence in the ethnographic record' in Martin, D.L. & Frayer, D.W. (eds) *Troubled Times: Violence and Warfare in the Past,* Amsterdam: Gordon & Breach, 1-20

ENCOP - Environment and Conflicts Project, n.d., Centre for Security Studies and Conflict Research, Swiss Institute of Technology Zurich (ETHZ) and the Swiss Peace Foundation, Berne http://www.fsk.ethz.ch/encop accessed 11/10/2002

Evans, C., 1988 'Monuments and analogy: the interpretation of causewayed enclosures' in Burgess, C., Topping, P., Mordant, C. & Maddison, M. (eds) 1988 *Enclosures and Defences in the Neolithic of Western Europe,* British Archaeological Reports International Series 403: 47-73

Ezzo, J.A., 1993 *Human Adaptation at Grasshopper Pueblo, Arizona: Social and Ecological Perspectives,* Ann Arbor, International Monographs in Prehistory

Ezzo J.A.& Johnson C.M. 1997, 'Analytical Perspectives on Prehistoric Migration: A Case Study from East-C Central Arizona' *Journal of Archaeological Science,* **24**(5), 447-466: Academic Press

Farwell, D.E., Andrews, P. & Brook, R., 1999 *Prehistoric, Roman and Early Saxon Settlement at Prospect Park, London Borough of Hillingdon,* Salisbury, Wessex Archaeology

Feinman, G.M., Lightfoot, K.G. & Upham, S., 2000 'Political Hierarchies and Organizational Strategies in the Puebloan Southwest' *American Antiquity, 6***5**(3), 449-470

Ferguson, R.B., 1997 'Violence and war in prehistory' in Martin, D.L. & Frayer, D.W. (eds), *Troubled Times: violence and war in the past,* Amsterdam: Gordon & Breach, 321-355

Ferrill, A., 1985 *The Origin of War: From the Stone Age to Alexander the Great,* London:Thames & Hudson

Field, D., 2001 'Place and Memory in Bronze Age Wessex', in J. Brück (ed)., *Bronze Age Landscapes: Transition and Transformation,* Oxford: Oxbow, 57- 64.

Filer, J., 1997, 'Ancient Egypt and Nubia as a source of information for cranial injuries' in J. Carman (ed) *Material Harm: archaeological studies of war and violence,* Glasgow: Cruithne Press, 47-74

Flannery, K.V., 1972 'The Cultural Evolution of Civilizations' *Annual Review of Ecology and Systematics,* 399-426

Flannery, K. & Marcus, J., 2003 'The origin of war: new ^{14}C dates from Ancient Mexico'. *P.N.A.S.,* www.pnas.org/cgi/doi/10.1073/pnas.1934526100, accessed 15/10/03

Ford, S., 1987 *East Berkshire Archaeological Survey,* Berkshire County Council Department of Highways and Planning Occasional Paper No.1

Ford, S. with Entwistle R & Taylor, K., 2003 *Excavations at Cippenham, Slough, Berkshire, 1995-7*, Reading: Berkshire Archaeological Society Monograph

Franzén, L. & Larsson, T.B., 1998 'Landscape Analysis and Stratigraphical and Geochemical Investigations of Playa and Alluvial Fan Sediments in Tunisia and Raised Bog Deposits in Sweden: A possible correlation between extreme climate events and cosmic activity during the late Holocene', in Peiser BJ, Palmer T & Bailey ME (eds) 1998 *Natural Catastrophes During Bronze Age Civilizations : Archaeological, geological, astronomical and cultural perspectives,* Oxford: British Archaeological Reports International Series 728, 148-161

Frayer, D.W., 1997 'Ofnet: Evidence for a Mesolithic Massacre' in Martin, D.L. & Frayer, D.W. (eds) *Troubled Times: Violence and Warfare in the Past,* Amsterdam: Gordon & Breach

Freestone, I.C., 1991 The petrography of the stone materials in Needham, S., 1991 *Excavation and Salvage at Runnymede Bridge 1978,* London: British Museum Press, 138-139

Gams, H. & Nordhagen, R., 1923 Postglaciale Klimaänderungen und Erdkrustenbewegungen in Mitteleuropa. *Landskundliche Forschungen* **25**.

Garlake, P.S., 1987 'Themes in the prehistoric art of Zimbabwe' *World Archaeology* **19**(2): 178-193

Gasparovic, F., Simoncic, V. & Subasic, D., 1995 'Croatian Environmental Problems Caused By War Opportunities for new remediation techniques and testing' in

Richardson M., *The Effects of War on the Environment: Croatia,* London: E & FN Spon,155-173
Glenny, M 1999 *The Balkans 1804-`999:Nationalism, War and the Great Powers,* London: Granta
Glob, P., 1973, *The Mound People: Danish Bronze-Age Man Preserved,* London: Book Club Associates Goodall, J., 1990 *Through a Window,* Harmondsworth, Penguin
Graves, M.A., Longacre, W.A. & Holbrook, S.J., 1982 'Aggregation and Abandonment at Grasshopper Pueblo, Arizona' *Journal of Field Archaeology* **9**(2), 193-206
Greenhill, B., 1995 *The Archaeology of Boats and Ships: an introduction,* London: Conway Maritime Press, 9
Greenwell, W. & Brewis, W.P., 1909 'The origin, evolution and classification of the bronze spearhead in Great Britain and Ireland' *Archaeologia* **61**: 439-472
Haas, J., 2001a 'Cultural Evolution and Political Centralisation' in Haas, J. (ed) *From Leaders to Rulers,* New York: Kluwer Academic/Plenum Press, 3-18
Haas, J., 2001b 'Warfare and the Evolution of Culture' in Feinman, G.M. & Price, T.D. (eds) *Archaeology at the Millennium: a sourcebook,* New York: Kluwer Academic/ Plenum, 329-350
Hamilton, E., 1995 'Was there ever a Roman Conquest?' in Hill, J.D. & Cumberpatch, C.G., (eds) *Different Iron Ages: Studies on the Iron Age in Temperate Europe,* Oxford: BAR Publishing, British Archaeological Reports International Series 602, 37- 44
Hanson, V.D., 1983 *Warfare and Agriculture in Classical Greece,* Pisa: Giardini
Harding, A. 1999 'Warfare: a Defining Characteristic of Bronze Age Europe?' in Carman, J. & Harding, A. (eds) *Ancient Warfare,* Stroud: Sutton, 157-173
Härke, H., 1990 Warrior Graves? The background of the Anglo-Saxon burial rite, *Past and Present,* **126**: 22-43
Haselgrove, C., 2002 'The Romanization of Belgic Gaul: some archaeological perspectives' in Blagg, T. & Millett, M., (eds) 2002 *The Early Roman Empire in the West* Oxford: Oxbow, pp 45-71
Haselgrove, C., 1996b 'La Romanisation de l'habitat rural dans la vallée de l'Aisne d'après les prospections de surface et les fouilles récentres', *Revue archéologique de Picardie, no. sp.* ***11,*** 109-120
Haselgrove, C., 1990a 'The Romanization of Belgic Gaul: some archaeological perspectives', in Blagg, T. & Millett, M. (eds) *The Early Roman Empire in the West,* Oxford: Oxbow, 45-71
Haselgrove, C., 1990b 'Later Iron Age Settlement in the Aisne Valley: some current problems and hypotheses', *Revue archéologique d'Ouest, Supp. no.****3,*** 249-259
Haselgrove, C., 1987 'Culture Process on the Periphery: Belgic Gaul and Rome during the late Republic and early Europe' in Rowlands, M., Larsen, M. & Kristiansen, K., (eds) *Culture and Periphery in the Ancient World,* Cambridge: Cambridge University Press, 104-123
Haselgrove, C., 1984 'Warfare and its Aftermath as reflected in the Precious Metal Coinage of Belgic Gaul', *Oxford Journal of Archaeology,* **3**(1): 81-105
Haselgrove, C., 2007 'The Age of Enclosure: Later Iron Age settlement and society in Northern France' in Haselgrove,C. & Moore, T. (eds) *The Later Iron Age in Britain and Beyond,* Oxford: Oxbow, 492-522
Haselgrove, C. & Scull, C., 1992 'The Romanization and de-Romanization of Belgic Gaul: the rural settlement evidence', in Wood, M. & Queiroga, F., (eds) *Current Research on the Romanization of the Western Provinces,* Oxford: BAR International Series S575, 9-23.
Hayman, G., 1998 Excavations at Home Farm, Laleham in 1997 *Surrey Archaeological Society Bulletin* No. 320, 2-3
Hedeager, L., 1987 'Empire, frontier and the barbarian hinterland: Rome and northern Europe from AD1- 400' in Rowlands, M., Larsen, M. & Kristiansen, K. (eds) *Centre and Periphery in the Ancient World* Cambridge: Cambridge University Press, 125-139
Hedges, R.E.M., Housley, R.A. Bronk, C.R. & Van Klinken, G.J., 1991 'Radiocarbon dates from the Oxford AMS system', *Archaeometry,* **33**: 121-134
Hedges, J. & Buckley, D., 1978 'Excavations at a Neolithic Causewayed Enclosure, Orsett, Essex, 1975' *Proceedings of the Prehistoric Society,* **44**: 219-308
Hegmon, M., 2005 'Beyond the Mold: Questions of Inequality in Southwest Villages', in: Pauketat, T.R. & Loren, D.D., (eds) *North American Archaeology,* Oxford: Blackwell, 212-234
Helbling, J., 2006 War and Peace in Societies without Central Power: Theories and Perspectives, in Otto, T., Thrane, H. & Vankilde, H. (eds) *Warfare and Society: Archaeological and Social Anthropological Perspectives,* Aarhus, Arrhus University Press, 113- 139
Hill, J.B., Clark, J.J., Doelle, W.H. & Lyons, P.D., 2004 'Prehistoric Demography in the Southwest: Migration, Coalescence, and Hohokam Population Decline', *American Antiquity,* **69**(4). 689-716
Hill, J.D., 2001 Hillforts and the Iron Age of Wessex, in Champion, T.C. & Collis, J.R. (eds) *The Iron Age in Britain and Ireland,* Sheffield: JRCollis, 95-116
Hill, J.N., 1972 'A Prehistoric Community in Eastern Arizona' in Leone, M.P.(ed) *Contemporary Archaeology: A Guide to Theory and Contributions,* Carbondale and Edwardsville, Southern Illinois University Press, 320-332
Hill, P. & Wileman, J., 2002 *Landscape of War*, Stroud: Tempus
Hingley, R., 1990 'Boundaries surrounding Iron Age and Romano-British settlements' *Scottish Archaeological Review* 7: 96-103
Holbrook, S.J., 1982 'The Prehistoric Local Environment of Grasshopper Pueblo, Arizona', *Journal of Field Archaeology* **9**(2), 207-215
Hodder. I. 1978 'Social organisation and human interaction: the development of some tentative hypotheses in terms of material culture', in Hodder., I. (ed) 1978 *The Spatial Organisation of Culture,* London: Duckworth, 199-269
Hodson, F.R., 1971 Three Iron Age brooches from Hammersmith in Sieveking G., (ed) *Prehistoric and Roman Studies*, London, British Museum Press, 50-57

Hollstein, E., 1980 Mitteleuropäische Eichendendrochronologie, Trier, *Grab. und Forsch* **11**, Mainz.

Jochim, M.A., 1981 *Strategies for Survival: Culture Behaviour in an Ecological Context,*New York and London: Academic Press

Johnson, B., 1975 *Archaeology and the M25*, Guildford: Surrey Archaeological Society Monograph

Joyce, A.A. & Johannessen, S. , 1993 'Abandonment and the production of archaeological variability at domestic sites' in Cameron, C.M. & Tomka, S.A., (Eds) *Abandonment of Settlements and Regions: Ethnoarchaeological and archaeological approaches,* Cambridge:Cambridge University Press, 138-153

Keegan. J. 1993 *A History of Warfare.* New York: Academic

Keeley, L.H., 1997 'Frontier warfare in the early Neolithic' in Martin, D.L. & Frayer, D.W. (eds), *Troubled Times: violence and war in the past,* Amsterdam: Gordon & Breach, 303-319

Keeley, L.H. 1996 *War Before Civilization* New York & Oxford: Oxford University Press

Keeley, L.H. & Cahen, D., 1989 'Early Neolithic Forts and Villages in NE Belgium: A Preliminary Report' *Journal of Field Archaeology,* **16**: 157-176

Klassen, M.A., 1998, 'Icon and narrative in transition: contact-period rock-art at Writing-on-Stone, southern Alberta, Canada' in Chippindale, C. & Taçon, P.S.C., *The Archaeology of Rock Art,* Cambridge: Cambridge University Press, 42-72

Krausse, D., 2003 Cultural Change and Settlement Development in the "Arduenna Silva" from 600 to 50BC, in in Bourgeois, J., Bourgeois, I. & Cherretté, B., (eds), *Bronze Age and Iron Age Communities in North-Western Europe,* Brussel: Koningluke Vlaamse Academie van Belgie voor Wetenschappen en Kunsten, 151-173

Kreuz, A., 1999 'Becoming a Roman Farmer: preliminary report on the environmental evidence from the Romanization project', in Creighton, J.D. & Wilson, R.J.A. (eds) *Roman Germany. Studies in Cultural interaction, Journal of Roman Archaeology Supp. Ser. no.* ***32,*** Portsmouth, Rhode Island, 71-98

Kristiansen, K., 2001 'Rulers and Warriors: Symbolic transmission and social transformation in Bronze Age Europe' in Haas, J. (ed) *From Leaders to Rulers,*New York: Kluwer Academic/Plenum Press, 85-104

Kristiansen, K., 1983 'Kriger og høvding i Danmarks Bronzealder. Et bidrag til bronzealdersvaerdets kulturhistorie' *Struktur och förändring i bronsålderns samhälle. Rapport från det tredje nordiska sumposiet för bronsåldersforskning i Lund, 23-25 april 1982,* Lund: Institute of Archaeology Report Series No. 17.

Kristiansen, K., 1982a The formation of tribal systems in later European prehistory: Northern Europe 4000-500 B.C., in Renfrew, C., Rowlands, M.J. & Segraves, B.A. (eds) *Theory and Explanation in Archaeology*, NY and London, 241-280

Kristiansen, K., 1982b 'Ekkehard Aner (†) und Karl Kerstein, Die Funde der älteren Bronzezeit des Nordischen Kreises in Dänemark, Schleswig-Holstein und Nieder-Sachsen. Band 1-5 (Besprechung)', *Germania* **60,** 2:590-600

Kristiansen, K., 1978 'The consumption of wealth in Bronze Age Denmark. A study in the dynamics of economic processes in tribal societies' in K.
Kristiansen & C. Paludan-Müller (eds), *New Directions in Scandinavian Archaeology,* Copenhagen: National Museum of Denmark

Kristiansen, K. & Rowlands, M., 1998 *Social Transformations in Archaeology: Global and Local Perspectives,* London & New York: Routledge

Krzisnik-Bukic, V, 1991 *Cazinska buna 1950,* Sarajevo, 3

Kuckelman, K.A., Lightfoot, R.R. & Martin, D.L., 2002 'The Bioarchaeology and Taphonomy at Castle Rock and Sand Canyon Pueblos, *American Antiquity* **67**(3): 486-513

Lamb, H.H., Lewis, R.P.W. & Woodroffe, A., 1966 Atmospheric circulation and the main climatic variables between 8000 and 0 BC: Meteorological evidence, in Sawyer, J.S. (ed) *Proceedings of the International Symposium on World Climate 8000 to 0 BP,* 174-217

Lambert, P.M., 1997 'Patterns of violence in prehistoric hunter-gatherer societies of Coastal Southern California' in Martin, D.L. & Frayer, D.W. (eds) *Troubled Times: Violence and Warfare in the Past,* War and Society Vol. 3, Amsterdam: OPA/Gordon and Breach, 77-110

Lambot, B., 1996 'Les Rèmes à la veille de la Romanisation. Le Porcien au 1er siècle avant J.-C.', *Revue archéologique de Picardie, no. sp.* ***11,*** 13-38

Lambrick, G., 1998 'Frontier Archaeology along the Thames', *British Archaeology*, April 1998, 12-13.
LeBlanc, S.A., 2001 'Warfare and Aggregation in the El Morro Valley, New Mexico' in Rice, G E & LeBlanc, S A (eds) *Deadly Landscapes - Case Studies In Prehistoric Southwestern Warfare,* Salt Lake City, University of Utah Press, 19-49

LeBlanc, S.A., 1999 *Prehistoric Warfare in the American Southwest,* Salt Lake City, University of Utah Press

LeBlanc, S. A, & Rice, G. E., 2001 'Southwestern Warfare - The Value of Case Studies' in Rice, G E & LeBlanc, S A (eds) *Deadly Landscapes - Case Studies In Prehistoric Southwestern Warfare,* Salt Lake City, University of Utah Press, 1-18

Lekson, S.H., 2005 'Chaco and Paquimé: Complexity, history, landscape' in Pauketat, T.R. & Loren, D.D. (eds) *North American Archaeology,* Oxford: Blackwell, 235-272

Lepetz, S., 1996 'Effets de la Romanisation sur l'élevage dans les établissements ruraux du Nord de la Gaule: l'exemple de l'augmentation de la stature des animaux domestiques' *Revue archéologique de Picardie, no. sp.* ***11,*** 317-324

L'Helgouac'h, J., 1998 'L'art mégalithique en Europe: Présentation générale' *L'Art des Mégalihes Peints et Gravés, Dossiers d'Archeologie* No. **230**, 4- 13

Lightfoot, R.R. & Kuckelman, K.A., 2001 "A Case of Warfare in the Mesa Verde Region" in Rice, G E &

LeBlanc, S A (eds) *Deadly Landscapes - Case Studies In Prehistoric Southwestern Warfare,* Salt Lake City, University of Utah Press, 51-64

Lobb, S.J. & Rose, P.G., 1996 *Archaeological Survey of the Lower Kennet Valley, Berkshire,* Salisbury: Wessex Archaeology Report No.9

Longley, D., 1976 The Archaeological Implications of Gravel Extraction in North-West Surrey, in *Research Volume of the Surrey Archaeological Society No. 3,* Old Woking: Surrey Archaeological Society, 1-35

Louwe Kooijmans, L.P., 1993 'An Early/ Middle Bronze Age multiple burial at Wassenaar, the Netherlands" *Analecta Praehistorica Leidensia* **26**:1-20

Lowell, J.C., 2007 'Women and Men in Warfare; implications of Gender Imbalance in the Grasshopper Region of Arizona', *American Antiquity* **72**(1): 95-123

Luckenbill, D., 1926 *Ancient Records of Assyria and Babylon,* New York, 87-88

Makkay, J., 2000 *An early war: The Late Neolithic mass grave from Esztetrgályhorváti,* Budapest, Makkay

Maloney, C., 1999 'Fieldwork roundup, 1998' in Maloney, C. & Holroyd, I., London fieldwork and publication round-up, 1998, *London Archaeologist,* **9,** supplement 1, 1-30

Maschner, H.D.G., 1997 'The evolution of Northwest Coast Warfare' in Martin, D.L. & Frayer, D.W. (eds) *Troubled Times: Violence and Warfare in the Past,* War and Society Vol. 3, Amsterdam: OPA/Gordon and Breach, 267-302

Maschner, H.D.G. & Patton, J.Q., 1997 'Kin selection and the origin of hereditary social inequality: a case study from the Northern Northwest Coast' in Maschner, H.D.G., (ed) *Darwinian Archaeologies,* New York and London: Plenum Press

Maschner, H.D.G. & Reedy-Maschner, K.L., 1998 'Raid, Retreat, Defend (Repeat): The archaeology and ethnohistory of warfare on the North Pacific Rim', *Journal of Anthropological Archaeology,* **17**: 19-51

McOmish, D., Field, D. & Brown, G., 2002 *The Field Archaeology of the Salisbury Plain Training Area,* Swindon, English Heritage

Megaw, R. & Megaw, V., 1989, *Celtic Art,* London: Thames & Hudson

Meggitt, M., 1977 *Blood is their argument,* Palo Alto, California: Mayfield

Meniel, P., 1996 'Les faunes des établissements ruraux de La Tène finale dans le Nord de la France', *Revue archéologique de Picardie, no. sp.* ***11,*** 309-316

Mercer, R.J., 1981, 'Excavations at Carn Brea, Illogan, Cornwall, 1970-73: A Neolithic Fortified Complex of the Third Millennium bc' reprinted as monograph from *Cornish Archaeology,* No.**20**

Mercer, R.J., 1988 'Hambledon Hill, Dorset, England' Hill' in Burgess, C., Topping, P., Mordant, C. & Maddison, M. (eds) *Enclosures and Defences in the Neolithic of Western Europe,* Oxford: British Archaeological Reports International Series 403: 89- 106

Mercer, R., 1989 The Earliest Defences in Western Europe: Part I Warfare in the Neolithic, *Fortress* **2**, 16-22

Mercer, R.J., 1999, 'The Origins of Warfare in the British Isles' in J. Carman & A. Harding, (eds), *Ancient Warfare: Archaeological Perspectives,* Stroud: Sutton, 143-156

Mertens, J., 1983 'The Military Origins of some Roman Settlements in Belgium' in Hartley, B. & Wacher, J. (eds) *Rome and her Northern Provinces,* Stroud: Sutton, 155-168

Milner, G.R., 1995 'An Osteological Perspective on Prehistoric Warfare' in Beck, L.A., (ed) *Regional Approaches to Mortuary Analysis,* New York & London: Plenum Press, p221-244

Milner, G.R., Anderson, E. & Smith, V.G., 1991 'Warfare in late Prehistoric West-Central Illinois" *American Antiquity* **56(**4): 581-602

MoLAS, 1996 *Annual Review for 1995,* London: Museum of London

Monks, S.J., 1997 Conflict and Competition in Spanish Prehistory - the role of warfare in societal development from the late fourth to third millennium BC, *Journal of Mediterranean Archaeology,* **10**(1): 3- 32

Montgomery, B.k. & Reid, J.J., 1990 'An Instance of Rapid Ceramic Change in the American Southwest, *American Antiquity,* **55:**88-97

Moore, J. & Jennings, D., 1992 *Reading Business Park: A Bronze Age Landscape,* Oxford: Oxford University Committee for Archaeology

Morphy, H., 1989 'Introduction' in Morphy, H. (ed), *Animals Into Art,* London: Unwin Hyman, 1-17

Namur: Division du Patrimoine, 1999 *Chronique de l'Archéologique Wallonne,* **7**

National Park Service, US Department of the Interior, 2004 *Tonto National Monument, Arizona*

Natural Environment Research Council/Institute of Geological Sciences, 1960 (3rd edn) *British Regional Survey: London and Thames Valley,* London:HMSO

Needham, S., 1991*Excavation and Salvage at Runnymede Bridge 1978,* London: British Museum Press

Needham, S., 1986 'The Metalwork - Discussion' in O'Connell, M., *Petters Sports Field, Egham: excavation of a Late Bronze Age/Early Iron Age site.* Guildford: Surrey Archaeological Society Research Volume No. 10, 57-60

Needham, S., 1979 'Two recent British shield finds and their continental parallels' *Proceedings of the Prehistoric Society,* **45**: 111-135

Needham, S. & Longley, D., 1980 'Runnymede Bridge, Egham. A Later Bronze Age Riverside Settlement' in Barrett, J. & Bradley, R., (eds) *Settlement and Society in the British Later Bronze Age,* Oxford: British Archaeological Reports 83 (ii): 397-436

Needham, S. & Spence, T., 1996 *Refuse and Disposal at Area 16 East Runnymede. Runnymede Bridge Research Excavations, Volume 2,* London: British Museum Press

Nelson, M.C. & Hegmon, N., 2001 'Abandonment is not as it seems: An approach to the relationship between site and regional abandonment', *American Antiquity* **66**(2):213-235

NERC/IGS., (National Environment Research Council/ Institute of Geological Science, 1960 *British Regional Geology. London and the Thames Valley* (3rd Edition). London: HMSO

Nordbladh, J. 1989 'Armour and fighting in the south Scandinavian Bronze Age, especially in view of rock art representation' in Larson, T.B. & Lundmark, H., *Approaches to Swedish Prehistory - A Spectrum of Problems and Perspectives in Contemporary Research,* Oxford: British Archaeological Reports International Series 500, 323-333

Northover, P., 1982 The exploration of long-distance movement of bronze in Bronze and Early Iron Age Europe, *Bulletin of the University of London Institute of Archaeology*, No. 19, 45-72

O'Connell, M., 1986 *Petters Sports Field, Egham: excavation of a Late Bronze Age/Early Iron Age site* Guildford: Surrey Archaeological Society Research Volume No. 10

Office of the High Representative (OHR), 1998, Reconstruction and Return Task Force, Herzegovina *RRTF: Report March 1998*, file://C:\WINNT\Profiles\edx130\Pers...\OHR Reconstruction and Return, March 1998.ht accessed 01/05/03

Ohlsson, L., 1999 *Environment Scarcity and Conflict: a study of Malthusian concerns,* Goteborg: Department of Peace and Development Research, Goteborg University

Okun, M.L., 1989 *The Early Roman Frontier in the Upper Rhine Area: Assimilation and Acculturation on a Roman Frontier,* Oxford: BAR International Series 547

Oliver, T.J., 2001 'Warfare in Tonto Basin' in Rice, G.E. & LeBlanc, S.A. (eds) *Deadly Landscapes: Case Studies in Prehistoric Southwestern Warfare,* Salt Lake City: University of Utah Press, 195-218

Oosterbeek, L., 1997 'War in the Chalcolithic? The Meaning of Western Mediterranean Chalcolithic Hillforts' in Carman, J. (ed) *Material Harm – archaeological studies of war and violence,* Glasgow: Cruithne Press 116-132

Osgood, R., 1999 'Britain in the age of warrior heroes' *British Archaeology,* 46 (July 1999): 8-9 Osgood, R., 1998 *Warfare in the Late Bronze Age of North Europe,* Oxford: British Archaeological Reports International Series 694

Osgood, R. & Monks, S. with Toms, J., 2000 *Bronze Age Warfare,* Stroud: Sutton O'Shea, J. & Halstead, P., 1989 'Conclusions: bad year economics' in Halstead, P. & O'Shea, J. (eds) *Bad Year Economics: cultural responses to risk and uncertainty,* Cambridge: Cambridge University Press, 123-126

Oswald, A., Dyer, C. & Barber, M. 2001 *The Creation of Monuments - Neolithic Causewayed Enclosures in the British Isles,* Swindon: English Heritage

Otterbein, K.F., 1989 *The Evolution of War - a cross-cultural study* 3rd ed., New Haven, Ct: HRAF Press

Pakenham, T., 1979 *The Boer War*, New York: Abacus

Pantreau, J.P., 1989 'The transition from Bronze Age to Iron Age in France: Economic, cultural and spiritual change' in Stig Sorensen, M.L. & Thomas, R. (eds) *The Bronze Age- Iron Age Transition in Europe – Aspects of Continuity and Change in European Societies c.1200-500 BC,* Oxford: British Archaeological Reports International Series **483(**I): 205-262

Parker-Brewis, W., 1923 'The bronze sword in Great Britain' *Archaeologia* **73**: 253-265

Parker Pearson, M., 1999 *The Archaeology of Death and Burial,* Stroud: Sutton

Parkinson, R., 1973 *The Peninsular War,* London: Hart-Davis MacGibbon

Parry, M. 1978 *Climate Change. Agriculture and Settlement,* Cambridge: Cambridge University Press for the Intergovernmental Panel on Climate Change

Pearson, J.L., 2002, *Shamanism and the Ancient Mind,* Walnut Creek: Alta Mira

Peebles, C.S. & Kus, S.M., 1977 'Some archaeological correlates of ranked societies" *American Antiquity* **42**(3): 421-447

Pettit, P., 2000 'Odd Man Out - Neanderthals and Modern Humans' *British Archaeology* **51**:8-13

Phillipson, D.A., 1974 'Iron Age history and archaeology in Zambia', *Journal of African History* **15**

Piggott, S., 1954 *The Neolithic Cultures of the British Isles,* Cambridge: Cambridge University Press

Price, T.D. Burton, J.H. & Bentley, R.A., 2002 'The Characterization of Biologically available Strontium Isotope Ratios for the study of Prehistoric Migration', *Archaeometry,* **44**(1), 117-135

Prilaux, G., 2000 'Une ferme gauloise spécialisée dans le travail du sel á Pont-Rémy "La Quete" et "Le Fond de Baraquin" (Somme). Évolution et particularités de l'époque en clos', *Revue Archéologique de Picardie, no. 1 / 2,* 233-254

Pritchard, J.B., 1969 *The Ancient Near East: Supplementary Texts and Pictures*, Princeton, Harvard University Press

Raftery, B., 1982 'Two recently discovered bronze shields from the Shannon Basin' *Journal of the Royal Society of Antiquaries of Ireland,* **112:** 5-17

Randsborg,K., 1999 'Into the Iron Age: a discourse on war and society' in Carman, J. & Harding, A. (eds) *Ancient Warfare,* Stroud: Sutton, 181-202

Randsborg, K., 1995, *Hjortspring - Warfare and Sacrifice in Early Europe*, Aarhus: Aarhus University Press

Ravesloot, J.C. & Spoerl, P.M., 1989 'The role of warfare in the development of status hierarchies at Casas Grandes, Chihuahua, Mexico' in Tkaczuk, D.C. & Vivian, B.C., (eds) *Cultures in Conflict: current archaeological perspectives, Proceedings of the Twentieth Annual (Chacmool) Conference of the Archaeological Association of the University of Calgary,* Calgary: University of Calgary, 7-18

Reddé, M., 1996 'Les Camps de Picardie' in Bénard, J. et al, *L'Armée Romaine en Gaule,* Paris: Editions Errance

Reddé, M. (ed), 1996 *L'Armée Romaine en Gaule,* Paris: Editions Errance

Reid, J.J., 1989 'A Grasshopper Perspective on the

Mogollon of the Arizona Mountains' in Cordell, L.S. & Gumerman, G.J. (eds) *Dynamics of Southwest Prehistory,* Washington DC & London: Smithsonian Institution Press, 65-96

Reid, J. & Whittlesey, S. 1997 *The Archaeology of Ancient Arizona,* Tucson: University of Arizona Press

Reid, J. & Whittlesey, S., 1999b *Grasshopper Pueblo,* Tucson: University of Arizona Press

Renfrew, C.,1973 'Monuments, Mobilisation and Social Organisation in Neolithic Wessex' in Renfrew, C. (ed) *The Explanation of Culture Change* London: Duckworth, 539-558

Rice, G.E., 2001 'Warfare and Massing in the Salt and Gila Basins of Central Arizona' in Rice, G.E. & LeBlanc, S.A. (eds) *Deadly Landscapes: Case Studies in rehistoric Southwestern Warfare,* Salt Lake City: University of Utah Press, 289-329

Rice, G E & LeBlanc, S A (eds) 2001 *Deadly Landscapes Case Studies In Prehistoric Southwestern Warfare,* Salt Lake City, University of Utah Press

Richardson, M., 1995 *The Effects of War on the Environment: Croatia,* London: E & FN Spon

Richmond, I.A., 1968 *Hod Hill, II,* London: British Museum Press

Roberts, C. & Manchester, K., 1995 *The Archaeology of Disease* (2nd ed)., Stroud: Sutton

Roper, M.K., 1975 'Evidence of warfare in the Near East from 10,000 - 4,300 BC' in Nettleship, M.A., Dale Givens, R. & Nettleship, A. (eds), *War, Its Causes and Correlates,* The Hague & Paris: Mouton, 299- 343

Rothschild, N.A., Mills, B.J., Ferguson, T.J. & Bublin, S. 1993 'Abandonment at Zuni farming villages' in Cameron, C.M. & Tomka, S.A., (Eds) *Abandonment of Settlements and Regions:Ethnoarchaeological and archaeological approaches,* Cambridge:Cambridge University Press, 123-137

Rowlands, M.J., 1972 'Defence: a factor in the organisation of settlements' in Ucko, P.J., Tringham, R. & Dimbleby, G.W., (eds)., *Man, Settlement and Urbanism,* London : Duckworth, 447-462

Roymans, N., 1990 *Tribal Societies in Northern Gaul: An anthropological perspective,* Amsterdam: Universiteit van Amsterdam

Savage, R.D.A., 1979 'Technical notes on the Watford sword fragments' Hampshire' in C.B. Burgess & D. Coombs, *Bronze Age Hoards,* Oxford: British Archaeological Reports British Series No. 67, 221-228

Schinkel, K., 2005 Hamlets on the move. Settlements in the southern and central parts of the Netherlands, in Louwe Kooijmanns, L.P., van den Broeke, E.W., Fokkens, H. & van Gijn, A.L. (eds), *The Prehistory of the Netherlands, Vol.2,* Amsterdam: Amsterdam University Press, 519-541

Schlanger, S.H. & Wilshusen, R.A.I., 1993 'Local abandonment and regional conditions in the North American Southwest', in Cameron, C.M. & Tomka, S.A., (Eds) *Abandonment of Settlements and Regions: Ethnoarchaeological and archaeological approaches,* Cambridge: Cambridge University Press, 85-98

Schmitt, S., 1998 *The Rwandan Genocide - Statistics of a Mass Grave - The Kibuye Case,* Http://garnet.acns.fsu.edu/~sss4407/Rwanda/RWStats .htm accessed 23rd May 2002

Selwyn, T., 2001 'Landscapes of Separation: Reflections on the symbolism of bypass roads in Palestine' in Bender, B &Winder, M (eds) *Contested Landscapes: Movement, Exile and Place,* Oxford and New York: Berg, 225-240

Shanks, M. & Hodder, I., 1995 'Processual. post- processual and interpretive archaeologies' in Hodder, I. et al (eds) *Interpreting Archaeology: Finding meaning in the past,* London:Routledge, 3-29

Sharples, N., 1991 'Warfare in the Iron Age of Wessex' *Scottish Archaeological Review* 8, 79-89

Sievers, S., 1996 'Armes celtiques, germaniques et romains: ce que nous apprenent les fouilles d'Alésia' in Reddé, M. (ed), *L'Armée Romaine en Gaule,* Paris: Editions Errance, 67-80

Simon, A.W. & Gosser, D.C., 2001 'Conflict and Exchange among the Salado of Tonto Basin: Warfare Motivation or Alleviation?' in Rice, G.E. & LeBlanc, S.A. (eds) *Deadly Landscapes: Case Studies in Prehistoric Southwestern Warfare,* Salt Lake City: University of Utah Press, 239-280

Smith, I.F., 1965 *Windmill Hill and Avebury:Excavations by Alexander Keiller 1925-1939',* Oxford: Clarendon Press

Smits, E. & Maat, G. 1993 'An Early/Middle Bronze Age common grave at Wassenaar, the Netherlands. The physical anthropological results' *Analecta Praehistorica Leidensia,* **26:** 21-28

Spoerl, P.M., 1984 'Prehistoric Fortifications in Central Arizona' in Spoerl, P.M. & Gumerman, G.J., (eds) *Prehistoric Cultural Development in Central Arizona: Archaeology of the Upper New River Region,* Carbondale: Southern Illinois University Centre for Archaeological Investigations Occasional Papers No. 5, 261-276

Spoerl, P.M., 1979 *Prehistoric Cultural Development and Conflict in Central Arizona,* PhD Dissertation, Carbondale: Southern Illinois University

Spratt , D.A., 1993 *Prehistoric and Roman Archaeology of North-East Yorkshire* Oxford: British Archaeological Reports British Series 104 :128

Stopford, J., 1987 'Danebury: an alternative view" *Scottish Archaeological Review,* **4** (Part 2): 70- 75

Tacitus (tr. H. Mattingley, 1948) *On Britain and Germany,* Harmondsworth: Penguin.

Taçon, P.S.C. & Chippindale, C., 1998 'An archaeology of rock-art through informed methods and formal methods' in Chippindale, C. & Taçon, P.S.C., *The Archaeology of Rock Art,* Cambridge: Cambridge University Press, 1-10

Thomas, R., 1998 'When land first became private property' *British Archaeology* **34** (May): 8-9

Thorpe, I.J., 2006 *Fighting and Feuding in Neolithic and Bronze Age Britain,* in Otto, T., Thrane, H. & Vankilde,

H., (eds) *Warfare and Society: Archaeological and Social Perspectives,* Aarhus: Aarhus University Press, 141-166

Tomka, S.A., 1993 'Site abandonment behaviour among transhumant agro-pastoralists: the effects of delayed curation on assemblage composition', in Cameron, C.M. & Tomka, S.A., (Eds) *Abandonment of Settlements and Regions: Ethnoarchaeological and archaeological approaches,* Cambridge:Cambridge University Press, 11-24

Tringham, R., 1972 'Territorial demarcation of prehistoric settlements' in Ucko, P.J., Tringham, R & Dimbleby, G.W., (eds)., *Man, Settlement and Urbanism,* London : Duckworth, 463-475

Tuggle, H.D. & Reid, J.J., 2001 "Conflict and Defense in the Grasshopper Region of East-Central Arizona" in Rice, G E & LeBlanc, S A (eds) *Deadly Landscapes - Case Studies In Prehistoric Southwestern Warfare,* Salt Lake City, University of Utah Press, 85-107

Turvalev, S. & Hajiyev, E., 1994 (tr. Hajiyev, E.) 'Impact of the War on the Environment' *Azerbaijan International* Autumn (2.3) ttp://www.azer.com/ai web/categories/ma.. Ider/23 warenvironment.htm accessed 11/10/2002

UN Truth Commission, 1992 *Report of Forensic Investigation, El Mozote, El Salvador 10 December 1992 to the Members of the United Nations Truth Commission.* Http://www.parascope.com/ articles/0197/el_mozdocb. htm accessed 23rd May 2002

Upham, S. & Plog, F., 1986 'The Interpretation of Prehistoric Political Complexity in the Central and Northern Southwest: Toward a Mending of the Models' *Journal of Field Archaeology,* **13**(2): 223- 238

van den Broeke, P., 2005 Gifts to the gods. Rites and cult sites in the Bronze Age and the Iron Age. in Louwe Kooijmanns, L.P., van den Broeke, E.W., Fokkens, H. & van Gijn, A.L. (eds), *The Prehistory of the Netherlands, Vol.2,* Amsterdam: Amsterdam University Press, 695-677

van der Dennen, J.M.G., (n.d.) *Ritualised 'Primitive' Warfare and Rituals in War:Phenocopy, homology, or...? Part 2: Ritual in 'Primitive' War - The function of human war ritual,* http://rint.rechten.rug.nl /rth/dennen/ritual2. htm, accessed 11/09/00

van der Dennen, J. & Falger, V. (eds), 1990 *Sociobiology and Conflict: evolutionary perspectives on competition, co-operation, violence and warfare,* London: Chapman & Hall

Van Geel, B., Raspopov, O.M., van der Plicht, J., & Renssen, H., 1998 'Solar Forcing of Abrupt Climate Change around 850 Calendar Years BC' in Peiser BJ, Palmer T & Bailey ME (eds) 1998 *Natural Catastrophes During Bronze Age Civilizations : Archaeological, geological, astronomical and cultural perspectives,* Oxford: British Archaeological Reports International Series 728, 162-168

van Hooff, J.A.R.A.M., 1990 'Intergroup competition and conflict in animals and man' in van der Dennen, J. & Falger, V. (eds) *Sociobiology and Conflict: Evolutionary perspectives on competition, co- operation, violence and warfare,* London: Chapman & Hall, 23-33

Van Lier, P., *Excavation of Mass Grave, Dos Erres, El Petén, June 1995,* http://www.peacebrigades.org/ guatemala/cap-piet05.html accessed 23rd May 2002

Van Lith, S.M.E., 1984 'Les Romains aux Pays-Bas et l'epigraphie' *Histoire et archéologie - lesDossiers* **86** ; 76-80

Vayda, A.P., 1968 'Primitive warfare' in Sills, D. (ed) *International Encyclopaedia of the Social Sciences,* **XVI**: 468-472

Vayda, A.P., 1967 'Maori warfare' in Bohannon, P. (ed) *Law and Warfare,* New York: Natural History Press, 359-380.

Vayda, A.P., 1960 *Maori Warfare,* Wellington: The Polynesian Society Vedder, H., 1966 *South West Africa in Early Times,* London:Sutton, 57-72

Vencl , S., 1999 'Stone Age Warfare' in Carman, J. & Harding, A. (eds) *Ancient Warfare: archaeological perspectives,* Stroud: Sutton, 57-72

Vencl, S., 1984 War and warfare in archaeology, *Journal of Anthropological Archaeology,* **3:** 116-132

Vermeulen, F. & Bourgeois, J., 2000 'Continuity of prehistoric burial sites in the Roman Landscape of Sandy Flanders', in Pearce, J., Millett, M. & Struck, M., (eds) *Burial, Society and Context in the Roman World,* Oxford: Oxbow, 142-161

von Clausewitz, C. 1976 [1832] (ed. and tr. M. Howard & P. Paret) *On War,* Princeton, NJ: Princeton University Press

Wahl, J. & König, H., 1987, 'Anthropologisch-traumalogische Untersuchung der menschlichen Skelettreste aus dem bandkeramischen Massengrab bei Talheim, Kreis Heilbronn', *Fundberichte aus Baden-Württemberg* **12:** 65-193

Wakely, J. 1997, 'Identification and Analysis of Violent and Non-violent Head Injuries in Osteo- archaeological Material' in Carman, J. (ed) *Material Harm: archaeological studies of War and Violence,* Glasgow: Cruithne Press, 24-46

Walker, P.L., 1989 'Cranial injuries as evidence of violence in prehistoric Southern California' *American Journal of Physical Anthropology,* **80**, 313- 323

Walthew, C.V., 1982 'Early Roman Town Development in Gallia Belgica: a review of some problems', *Oxford Journal of Archaeology* **1**(2): 225-236

Waters, M.R. & Ravesloot, J.C., 2001 'Landscape change and the cultural evolution of the Hohokam along the Middle Gila River and other river valleys in South-Central Arizona' *American Antiquity* **66**(2): 285-299

Webster, D., 1975 'Warfare and the evolution of the state: a reconsideration' *American Antiquity* **40**(4):464-470

Wellington, I., 2001 'Considering Continuity of Deposition on Votive Sites in Northeastern France from 200 BC to AD 100' in Carruthers, M.et al, *TRAC 2001 Proceedings of the 11th Annual Roman Archaeology, Glasgow,* Oxford: Oxbow, 1-12

Wells, C., 1964, *Bones, Bodies and Disease,* London: Thames & Hudson

Wenham, S.J., 1989 'Anatomical Interpretations of Anglo-

Saxon Weapon Injuries' p123-129 in Hawkes, S.R. (ed) *Weapons and Warfare in Anglo-Saxon England*, Oxford: Oxford University Committee for Archaeology Monograph 21, 123-129.

Wheeler, R.E.M., 1943 *Maiden Castle, Dorset*, Res.Rep. Comm.Soc. Antiq., **12**, Oxford

Whittle, A., 1996 *Europe in the Neolithic,* Cambridge: Cambridge University Press

Wigg, D.G. , 1999, 'The development of the monetary economy in N. Gaul in the late La Tene and early Roman periods' in Creighton, J.D. & Wilson, R.J.A. (eds) *Roman Germany. Studies in Cultural Interaction, Journal of Roman Archaeology Supp. Ser. no.* ***32,*** Portsmouth, Rhode Island, 99-124

Wightman, E.M., 1985 *Gallia Belgica,* London: Batsford

Wightman, E.M., 1977 'Military Arrangements, Native Settlements and related developments in Early Roman Gaul', *Helinium,* **XVII**: 105-126

Wightman, E.M., 1974 'La Gaule Chevelue entre César et Auguste', *Actes de 9ième congrès d'études sur les frontières romaines, Mamaïa 1972,* 474-5

Wilcox, D.R., Robertson, G. Jr., & Wood, J.S. 2001 'Antecedents to Perry Mesa: Early Pueblo III Defensive Refuge Systems in West-Central Arizona' and 'Organized for War: ThePerry Mesa Settlement System and its Central Arizona Neighbors' in Rice, G.E. & LeBlanc, S.A. (eds) *DeadlyLandscapes: Case Studies in Prehistoric Southwestern Warfare,* Salt Lake City:University of Utah Press, 109-140 and 141-194

Wilson, D.R., 1984 'Forts et Camps Romains Vus de Ciel' *Histoire et archéologie - les Dossiers* **86** : 65-70

Wood, J.S., 2000 'Vale of Tiers Palimpsest: Salado Settlement and Internal Relationships in the Tonto Basin Area' in Dean, J.S. (ed) *Salado*, Amerind Foundation, 107-141

Wright, Q., 1962, *A Study of War,* Chicago Yates, D.T., 2007 *Land, Power and Prestige. Bronze Age Field Systems in Southern England,* Oxford: Oxbow Books

Yates, D.T., 2001 'Bronze Age Agricultural Intensification in the Thames Valley and Estuary, in Brück, J. (ed) *Bronze Age Landscapes. Tradition and Transformation,* Oxford: Oxbow Books, 65-83

Yates, D.T., 1999 'Bronze Age Field Systems in the Thames Valley', *Oxford Journal of Archaeology* **18**(2), 157-170

York, J., 2002 'The Life Cycle of Bronze Age Metalwork from the Thames', *Oxford Journal of Archaeology* **21**(1), 77-92

Zimmerman, L.J. 1997 'The Crow Creek Massacre: Archaeology and Prehistoric Plains Warfare in Contemporary Contexts' in J.Carman (ed) *Material Harm: archaeological studies of war and violence,* Glasgow: Cruithne Press, 75-94

www.ingramcontent.com/pod-product-compliance
Lightning Source LLC
LaVergne TN
LVHW071212210926
848482LV00015B/792